Mathematics Education – An Asian Perspective

Series editors

Berinderjeet Kaur, National Institute of Education, Singapore, Singapore
Catherine Vistro-Yu, Ateneo de Manila University, Manila, Philippines

Aims and Scope

Mathematics Education – An Asian Perspective facilitates high quality publications on rigorous aspects of mathematics education in Asia. This will be achieved by producing thematic books that capture knowledge and practices on mathematics education in Asia from both the insider and outsider perspectives. The series helps to establish a much needed Asian perspective to mathematics education research in the international landscape.

Over the last decade or so, several international comparative studies have shed light on systems of schooling that were otherwise not very much sought after. Several educational systems in Asia, in particular East Asia have consistently produced stellar outcomes for mathematics in both TIMSS and PISA despite the fact that both studies measure achievement in mathematics in distinct ways that are very much orthogonal to each other, while other Asian systems have not been able to replicate the same level of success. Though one may occasionally chance upon a publication on some aspect of mathematics education in Asia, there appears to be in general a dearth of publications on mathematics education in Asian countries from the perspectives of scholars from Asia. Hence it is apparent that there is a gap in the availability of knowledge on mathematics education from the region in the international space.

This series has a wide scope with emphasis on relevancy and timeliness. It encompasses the general trends in educational research such as theory, practice and policy. Books in the series are thematic and focus both on macro and micro topics. An example of a themed book on a macro topic could be one on "School mathematic curricula – An Asian perspective" while a themed book on a micro topic could be one on "The pedagogy of 'simultaneous equations' in Asian classrooms".

More information about this series at http://www.springer.com/series/11679

Berinderjeet Kaur · Oh Nam Kwon
Yew Hoong Leong
Editors

Professional Development of Mathematics Teachers

An Asian Perspective

Springer

Editors
Berinderjeet Kaur
National Institute of Education
Nanyang Technological University
Singapore
Singapore

Yew Hoong Leong
National Institute of Education
Nanyang Technological University
Singapore
Singapore

Oh Nam Kwon
Seoul National University
Seoul
Korea, Republic of (South Korea)

ISSN 2366-0155 ISSN 2366-0163 (electronic)
Mathematics Education – An Asian Perspective
ISBN 978-981-10-2596-9 ISBN 978-981-10-2598-3 (eBook)
DOI 10.1007/978-981-10-2598-3

Library of Congress Control Number: 2016951660

Printed on acid-free paper

This Springer imprint is published by Springer Nature
The registered company is Springer Nature Singapore Pte Ltd.
The registered company address is: 152 Beach Road, #22-06/08 Gateway East, Singapore 189721, Singapore

Contents

1 **Mathematics Teacher Professional Development: An Asian Perspective** 1
Yew Hoong Leong, Berinderjeet Kaur and Oh Nam Kwon

Part I Policies, Structures, Frameworks and Contexts

2 **Professional Development of Secondary Mathematics Teachers in Mainland China** 17
Rongjin Huang, Lijun Ye and Kyle Prince

3 **The Endless Long-Term Program of Mathematics Teacher Professional Development in Indonesia** 33
Yaya S. Kusumah and Farida Nurhasanah

4 **Lesson Study: The Fundamental Driver for Mathematics Teacher Development in Japan** 47
Akihiko Takahashi

5 **Towards Balancing Knowledge and Practice of In-Service Mathematics Teacher Education Program in Korea** 63
Oh Nam Kwon, Jung Sook Park, Jaehee Park and Jee Hyun Park

6 **Profiling Mathematics Teacher Professional Development in Malaysia** 77
Chin Mon Chiew and Chap Sam Lim

7 **Congruence Between Context and Opportunities for Professional Development of Mathematics Teachers in the Philippines** 89
Debbie Verzosa, Maria Theresa Tulao-Fernando and Catherine Vistro-Yu

8 **Professional Development of Mathematics Teachers in Singapore** 97
Berinderjeet Kaur and Lai Fong Wong

9 Mathematics Teachers Professional Development in Taiwan 109
Fou-Lai Lin and Yu-Ping Chang

Part II Innovative Professional Development Programs in Asia

10 Constraints and Affordances in Bringing About Shifts in Practice Towards Developing Reasoning in Mathematics: A Case Study 121
Ruchi S. Kumar and K. Subramaniam

11 A Community-Based, Practice-Oriented Teacher Professional Development Program: Changes in Teaching Culture in Korea 141
Oh Nam Kwon, Jee Hyun Park, Jung Sook Park and Jaehee Park

12 Classroom as a Site for Teacher Learning: Emergence of a Paradigm Shift in Mathematics Teacher Education in Pakistan.... 159
Anjum Halai

13 Teaching for Metacognition Project: Construction of Knowledge by Mathematics Teachers Working and Learning Collaboratively in Multitier Communities of Practice........................ 169
Berinderjeet Kaur, Divya Bhardwaj and Lai Fong Wong

14 Boundary Objects Within a Replacement Unit Strategy for Mathematics Teacher Development 189
Yew Hoong Leong, Eng Guan Tay, Tin Lam Toh, Romina Ann Soon Yap, Pee Choon Toh, Khiok Seng Quek and Jaguthsing Dindyal

15 Facilitating Professional Growth of Taiwanese In-service Mathematics Teachers Through an Innovative School-Based Program 209
Fou-Lai Lin, Hui-Yu Hsu and Jian-Cheng Chen

16 School-Based In-service Mathematics Teachers' Professional Development: Designing Diagnostic Conjecturing Activities....... 223
Jian-Cheng Chen and Fou-Lai Lin

Editors and Contributors

About the Editors

Berinderjeet Kaur is Professor of Mathematics Education at the National Institute of Education in Singapore. Her primary research interests are in the area of classroom pedagogy of mathematics teachers and comparative studies in mathematics education. She has been involved in numerous international studies of mathematics education and was the Mathematics Consultant to TIMSS 2011. She is also a member of the MEG (Mathematics Expert Group) for PISA 2015. As the President of the Association of Mathematics Educators (AME) from 2004–2010, she has also been actively involved in the professional development of mathematics teachers in Singapore and is the founding chairperson of Mathematics Teachers' Conferences that started in 2005 and the founding editor of the AME Yearbook series that started in 2009. She was awarded the Public Administration Medal by the President of Singapore in 2006.

Oh Nam Kwon is Professor of Mathematics Education at Seoul National University. She has worked as a researcher, primary and secondary teacher educator, supervisor of graduate research and as an adviser to governments. She has written practically oriented books and articles for mathematics educators as well as producing a large set of research articles. Professor Kwon's research interests include curriculum development, teaching and learning of collegiate mathematics, teacher education, and inquiry-oriented teaching. She is Editorial Board for book series "Advances in Mathematics Education" by Springer. She has served as committee member for numerous international (including International Programme Committee of ICME-12) and Korean organizations of mathematics education. She is serving as a member of National Committee of Korean Institute of Curriculum and Evaluation. She was a member of 2015 OECD/PISA Mathematics Expert Group. She received the Best Teaching Awards, Seoul National University in 2009. She has served the Jury for Mathematics Planet Earth 2013 Virtual Modules Competition the patronage of UNESCO and the committee member for Leelavati Prize 2014 of International Mathematics Union. She received the Outstanding

Services Achievement Awards, Korean Society of Mathematical Education in 2014. Currently, she is the Chair of the 2015 Revised National High School Mathematics Curriculum in Korea.

Yew Hoong Leong is Assistant Professor with the Mathematics and Mathematics Education Academic Group, National Institute of Education, Nanyang Technological University in Singapore. He engages in design research that places emphasis on mathematics teacher development, problem solving, and teaching for low achievers.

Contributors

Divya Bhardwaj is research associate at the Mathematics and Mathematics Education Academic Group of the National Institute of Education (NIE), Singapore. Since 2011, she has been involved with various research projects at the Centre for Research in Pedagogy and Practice and the Mathematics and Mathematics Education Academic group in NIE. Prior to that, she was a secondary school mathematics teacher at NPS International School, Singapore. She completed her Master of Science (Mathematics for Educators) in NIE in 2009 and the Cambridge International Diploma for Teachers & Trainers in 2010. Currently, she is pursuing a Doctor of Philosophy in Mathematics.

Yu-Ping Chang is a Postdoctoral researcher at the Department of Mathematics of the National Taiwan Normal University. Her research interests involve mathematics textbook analysis, mathematics teacher education, and professional development.

Jian-Cheng Chen is Assistant Professor in Center for General Education of Ming Chi University of Technology. His research interests include students' mathematical thinking, mathematics teachers' professional development and mathematics teacher educators' professional development.

Chin Mon Chiew is Senior Lecturer in the Mathematics Department of Institute of Teacher Education Tuanku Bainun Campus, Penang, Malaysia. He obtained his Ph.D. in Mathematics Education from the Universiti Sains Malaysia, Malaysia. He teaches mathematics and mathematics education courses to pre-service mathematics teachers for primary schools. His research interests concern lesson study and professional learning communities, mainly in the area of teacher development. He has been actively involved as a master trainer to promote lesson study for teachers' professional development both at state and national level. At international involvement, he has presented several research papers in the EARCOME and ICME conferences.

Jaguthsing Dindyal is Associate Professor at the National Institute of Education, Nanyang Technological University in Singapore. He holds Ph.D. in Mathematics Education from Illinois State University, USA. He has prior experience of teaching at the secondary level and currently teaches mathematics education courses to both pre-service and in-service teachers. His research interests include reasoning and proofs, problem solving, geometric and algebraic thinking, international studies and the mathematics curriculum. His contributions in these areas include a number of refereed conference papers, book chapters and journal articles.

Anjum Halai is Professor of Mathematics Education currently based at the Aga Khan University in Pakistan. She has a wide experience of working in the field especially in the context of developing countries such as Pakistan, Kenya, Tanzania and Uganda. She is an active participant and contributor in international events such as the International Congress in Mathematics Education and the PME conferences. Her research interests are focussed on issues of social justice in multilingual mathematics classrooms. Her recent publications include the edited book 'Teaching and Learning Mathematics in Multilingual Classrooms: Issues for policy, practice and teacher education'.

Hui-Yu Hsu is Assistant Professor in Graduate Institution of Mathematics and Science Education at National Hsinchu University of Education. Her research interests include teaching and learning of geometry, mathematics textbook analysis, and mathematics teacher education.

Rongjin Huang is Associate Professor of Mathematics Education at Middle Tennessee State University, USA. His research interests include mathematic classroom instruction, mathematics teacher education, and comparative study in mathematics education. Dr. Huang has published scholarly work extensively. His recently published books include *How Chinese teach mathematics and improve teaching* (2013) and *Prospective mathematics teachers ' knowledge of algebra: A comparative study in China and the United States of America* (2014). Dr. Huang has served as a guest editor for *The International Journal on Mathematics Education*. He has organized and chaired secessions at AERA, NCTM, as well as ICME.

Ruchi S. Kumar is Assistant professor at Centre for Education Innovation and Action Research (CEIAR) at Tata Institute of Social Sciences, Mumbai, India. She is pursuing her Ph.D. in Mathematics Education from Homi Bhabha Centre for Science education, Tata Institute of Fundamental Research. Her research interests include analyzing interaction between teachers' beliefs, knowledge and practices, role of specialized content knowledge and pedagogical content knowledge in teaching of mathematics and role of communities in teachers' learning.

Yaya S. Kusumah is Professor of Mathematics Education at the Indonesia University of Education. He has Ph.D. in Mathematics from Curtin University Australia. He is currently the Director of Post Graduate Schools at the University. He has worked extensively in the area of mathematics and mathematics education. He has been involved in Lesson Study programs in Indonesia since 2005. In 2005 he became a member of university accreditation board in Indonesia as a reviewer. His research interests include the areas of discrete mathematics, professional development of teachers and technology for teaching mathematics.

Chap Sam Lim is Professor in Mathematics Education at the Universiti Sains Malaysia, Penang, Malaysia. She obtained Ph.D. in Mathematics Education from the University of Exeter, United Kingdom. She is currently teaching mathematics education research course and qualitative research method courses at graduate level. Her research interests focus on public images of mathematics, teaching mathematics in second language, cross-cultural study and particularly Lesson Study as a professional development for mathematics teachers. She is an active collaborator of several international research projects especially on Lesson Study and cross-cultural comparative study of mathematics teaching and learning in schools. Her latest publication includes as one of the co-editors in The First Sourcebook on Asian Research in Mathematics Education: China, Korea, Singapore, Japan, Malaysia and India published in 2015.

Fou-Lai Lin is Chair Professor in Mathematics Department at National Taiwan Normal University and the Director of the Shi-Da Institute for Mathematics Education where he directs a national program "Just Do Math" for students, teachers and teacher educators. He is the Founding Editor-in-Chief of International Journal of Science and Mathematics Education. He was a former president of PME. About 35 years engaging in mathematics education research and practice, his works across curriculum and material development, students' mathematics understanding, teachers' and teacher educators' education and professional development.

Farida Nurhasanah is Lecturer at the Teachers Training and Education Faculty of Sebelas Maret University in Indonesia. At present she is a doctorate student in Indonesia University of Education. Her research interests are abstraction in mathematics, professional development for in-service and pre-service mathematics teachers, problem solving, geometrical thinking.

Jaehee Park is a mathematics teacher at Gyeonggi Science High School for the Gifted in Korea, where she has taught since 2011. Also she has taught mathematics at several middle schools and high schools since 1993 and gave lectures at Silla University and Hanyang University for 4 years. She completed the doctoral coursework in mathematics education at Seoul National University. Her major areas of research are STEAM education focused on mathematics, student-centered instruction, and mathematics class based on questioning and discussion, etc.

Jee Hyun Park has a Ph.D. in Mathematics Education from Seoul National University. She has taught middle and high school mathematics in Seoul, Korea for nearly 17 years. She has also taught future teachers at university for 6 years. For the past 4 years, she has been on the research supported by Korea Foundation for Advance of Science and Creativity. She has been working on programs that develop secondary teachers' professionalism for over 10 years as teaching consultant. Her professional interests are finding ways to increase mathematics literacy for secondary students and improving professional development, especially teaching and assessing using the technology.

Jung Sook Park has a Ph.D. in Mathematics Education from Seoul National University. She has taught middle and high school mathematics in Seoul, Korea for nearly 23 years. She has also taught future mathematics teachers at Hanyang University for 8 years. She has attended in developing in revised 2015 National Standard Curriculum and high school textbooks. She has been working on programs that develop secondary teachers' professionalism.

Kyle Prince teaches mathematics at Central Magnet School in Murfreesboro, Tennessee, USA. Prince was named the 2015–2016 Middle Tennessee Teacher of the Year and selected as a finalist for the Presidential Award for Excellence in Science and Mathematics Teaching. He is currently pursuing his Ph.D. in Mathematics Education at Middle Tennessee State University. His interests include teaching through problem solving and adapting Chinese and Japanese lesson study models in the United States. In his dissertation, Prince is exploring lesson study as an aid in enhancing teachers' understandings, implementations, and perceptions of mathematics teaching practices.

Khiok Seng Quek is Senior Lecturer with the Psychological Studies Academic Group, National Institute of Education, Nanyang Technological University, Singapore. His areas of study focus on assessment and teacher education. He is an active collaborator on projects that investigate the integration, influence, and use of new assessment strategies in mathematics classrooms, including the recent Mathematics Assessment Project funded by the Centre for Research in Pedagogy and Practice, NIE. He is also the co-principal investigator of the project Mathematical Problem Solving for Everyone (M-ProSE).

K. Subramaniam (Ravi) is Professor of mathematics education at the Homi Bhabha Centre for Science Education, Tata Institute of Fundamental Research in Mumbai, India. His research has focused on the learning of core topics in middle school mathematics such as fractions, ratio and proportion, algebra and geometric measurement, specialized knowledge and professional development of in-service mathematics teachers and connecting out-of-school mathematical knowledge with school learning.

Akihiko Takahashi is Associate Professor at DePaul University. He teaches mathematics and mathematics education for prospective teachers. He was a teacher in Japan before becoming an educator of mathematics teachers. During his teaching career, he was nationally active in mathematics lesson study in Japan. He received his Ph.D. from the University of Illinois at Urbana-Champaign; his dissertation research focused on Internet use in mathematics education. He has published over 45 journal articles and book chapters and given over 50 presentations and keynote at conferences and workshops in Canada, Chile, Indonesia, Japan, Korea, Malawi, Malaysia, Mexico, Philippine, Qatar, Singapore, Thailand, Uganda, United Kingdom, and United States.

Eng Guan Tay is Associate Professor in the Mathematics and Mathematics Education Academic Group at the National Institute of Education, Singapore. His main research areas are in Graph Theory for Mathematics, and problem solving as well as teaching undergraduate mathematics for Mathematics Education.

Pee Choon Toh received his Ph.D. from the National University of Singapore in 2007. He is currently Assistant Professor at the National Institute of Education, Nanyang Technological University in Singapore. His main research interest includes number theory, specifically the theory of partitions, modular forms and elliptic functions. He is also interested in problem solving, the teaching of mathematics at the undergraduate level, as well as the use of technology in teaching.

Tin Lam Toh is Associate Professor with the Mathematics and Mathematics Education Academic Group, National Institute of Education, Nanyang Technological University, Singapore. He obtained his Ph.D. in Mathematics (Henstock-stochastic integral) from the National University of Singapore. Dr. Toh continues to do research in mathematics as well as in mathematics education. He has papers published in international scientific journals in both areas. Dr. Toh has taught in junior college in Singapore and was head of the mathematics department at the junior college level before he joined the National Institute of Education.

Maria Theresa Tulao-Fernando is a full-time instructor at the Mathematics Department of the Ateneo de Manila University. Her scholarly interests include teacher education and professional development, links from arithmetic to algebraic thinking, and mathematical pedagogical knowledge.

Debbie Verzosa is Assistant Professor in the Mathematics Department of the Ateneo de Manila University. Her main fields of interest are in the areas of non-formal education for disadvantaged groups, the development of children's mathematical thinking, and mathematics teacher education.

Catherine Vistro-Yu Professor in the Mathematics Department of the Ateneo de Manila University, the Philippines, is a mathematics educator. She teaches mathematics and mathematics education courses to classroom teachers of both the primary and secondary levels. Her research interests lie mostly in the area of mathematics teacher education and children's understanding of mathematics although she has studied other important concerns as well, such as the use of technology in mathematics education and mathematics teachers' beliefs. In the last decade, she has actively engaged in large-scale mathematics education projects and programs in her country that address issues of curriculum, teacher competencies in mathematics, and student achievement, among others. Catherine's international network has provided her with valuable opportunities for significant collaboration with colleagues in Asia through the SEACME and EARCOME, and with other foreign colleagues through ICME.

Lai Fong Wong has been a mathematics teacher for over 20 years and is known for her efforts in engaging students with fresh and creative strategies in the study of mathematics. For her exemplary teaching and conduct she was given the President's Award for Teachers in 2009. She sets the tone for teaching the subject in Anderson Secondary School, being Head of Department (Mathematics) from 2001 to 2009 and currently a Lead Teacher for Mathematics. She is also an executive committee member of the Singapore's Association of Mathematics Educators, and a member of the Singapore Academy of Teacher's Math Chapter Core Team. She was awarded the Post-graduate Scholarship by Singapore Ministry of Education to pursue a Master of Education in Mathematics which she has completed in 2014. Currently, she is involved in several Networked Learning Communities looking at ways to infuse mathematical reasoning, metacognitive strategies, and real-life context in the teaching of mathematics. Lai Fong is also active in the professional development of mathematics teachers and in recognition of her significant contribution toward the PD of Singapore teachers, she is awarded the Associate of the Academy of Singapore Teachers in 2015.

Romina Ann Soon Yap holds an M.S. degree in Applied Mathematics from the University of the Philippines and is currently completing her Ph.D. in Mathematics Education at the National Institute of Education, Nanyang Technological University in Singapore. She has worked as a secondary mathematics teacher in the Philippines and as a mathematics instructor at the Ateneo de manila University. Her research interests include mathematical problem solving and mathematics teacher development.

Lijun Ye is Professor of Mathematics Education at Hangzhou Normal University, China. His research interests include mathematics classroom research, mathematics teacher education, and comparative mathematics education. He has completed several Ministry of Education of China-funded research projects and published scholarly work extensively. He has published 80 papers and more than 20 textbooks. The recently published books include Mathematics Curriculum and Instruction (2011), Studies on Classroom Teaching of Mathematics (2014), and Teaching Design of Secondary School Mathematics (2015). He has organized and chaired activities at various international conferences such as AERA and ICME.

Contributors

Divya Bhardwaj National Institute of Education, Nanyang Technological University, Singapore, Singapore

Yu-Ping Chang National Taiwan Normal University, Taipei, Taiwan

Jian-Cheng Chen Ming Chi University of Technology, New Taipei City, Taiwan

Chin Mon Chiew Institute of Teacher Education, Penang, Malaysia

Jaguthsing Dindyal National Institute of Education, Nanyang Technological University, Singapore, Singapore

Anjum Halai Institute for Educational Development, Aga Khan University, Karachi, Pakistan

Hui-Yu Hsu National Hsinchu University of Education, Hsinchu City, Taiwan

Rongjin Huang Middle Tennessee State University, Murfreesboro, USA

Berinderjeet Kaur National Institute of Education, Nanyang Technological University, Singapore, Singapore

Ruchi S. Kumar Homi Bhabha Centre for Science Education, TIFR, Mumbai, India

Yaya S. Kusumah Department of Mathematics Education, Universitas Pendidikan Indonesia, Bandung, Indonesia

Oh Nam Kwon Seoul National University, Seoul, South Korea

Yew Hoong Leong National Institute of Education, Nanyang Technological University, Singapore, Singapore

Chap Sam Lim Universiti Sains Malaysia, Penang, Malaysia

Fou-Lai Lin National Taiwan Normal University, Taipei City, Taiwan

Farida Nurhasanah Mathematics Education Department of Teachers Training and Education Faculty, Sebelas Maret University, Surakarta, Indonesia

Jaehee Park Gyeonggi Science High School for the Gifted, Suwon, South Korea

Jee Hyun Park Banpo High School, Seoul, South Korea

Jung Sook Park Yangjae High School, Hanyang University, Seoul, South Korea

Kyle Prince Middle Tennessee State University, Murfreesboro, USA

Khiok Seng Quek National Institute of Education, Nanyang Technological University, Singapore, Singapore

K. Subramaniam Homi Bhabha Centre for Science Education, TIFR, Mumbai, India

Akihiko Takahashi DePaul University, College of Education, Chicago, USA; Project IMPULS at Tokyo Gakugei University, Tokyo, Japan

Eng Guan Tay National Institute of Education, Nanyang Technological University, Singapore, Singapore

Pee Choon Toh National Institute of Education, Nanyang Technological University, Singapore, Singapore

Tin Lam Toh National Institute of Education, Nanyang Technological University, Singapore, Singapore

Maria Theresa Tulao-Fernando Ateneo de Manila University, Quezon City, Philippines

Debbie Verzosa Ateneo de Davao University, Davao City, Philippines

Catherine Vistro-Yu Ateneo de Manila University, Quezon City, Philippines

Lai Fong Wong Anderson Secondary School, Singapore, Singapore

Romina Ann Soon Yap National Institute of Education, Nanyang Technological University, Singapore, Singapore

Lijun Ye Hangzhou Normal University, Hangzhou, China

Series Editors' Introduction

The second volume of the book series Mathematics Education: An Asian Perspective, entitled, "Professional Development of Mathematics Teachers: An Asian Perspective" and edited by Berinderjeet Kaur, Oh Nam Kwon and Leong Yew Hoong offers a counterpart to the extensive corpus of literature on the same topic that has been dominated by scholars from the Western countries. This volume is a worthy contribution to mathematics education literature especially relevant to Asian teachers and students. As envisioned, it provides a rich source of information and analyses that could only be rightfully written by scholars from Asia for scholars in Asia.

The book is the first to present substantial contributions from scholars in Asia in the area of professional development of mathematics teachers in their respective countries. A synthesis of the contributions by the editors sheds valuable light on how approaches to the professional development of mathematics teachers uniquely resemble or differ from those in the West. The book also provides scholars from non-English-speaking and underrepresented Asian countries the opportunity to engage in discourse with other scholars in the field.

There is no doubt that this book contributes towards narrowing the gap in the availability of knowledge on the development of mathematics teachers in Asia in the international space. We hope the readers will find it enjoyable and the researchers a valuable resource.

Singapore — Berinderjeet Kaur
Philippines — Catherine Vistro-Yu

Chapter 1
Mathematics Teacher Professional Development: An Asian Perspective

Yew Hoong Leong, Berinderjeet Kaur and Oh Nam Kwon

Abstract This introductory chapter sets the context for the book. It also provides an overview of the chapters in the two parts of the book. The first part comprises eight chapters on policies, structures, frameworks, and contexts. The chapters provide us with some ideas about the professional development (PD) of mathematics teachers in eight Asian countries, namely China, Indonesia, Japan, Korea, Malaysia, the Philippines, Singapore, and Taiwan. It is evident from these eight chapters that the countries are at different phases of development of teacher professionalism. In some countries there are mandatory acts and regulations governing the continuous teacher PD while in others the situation is lax, and in such cases PD would greatly depend on a teacher's own motivation and also the availability of resources. The second part comprises eight chapters that showcase innovative approaches to mathematics teacher PD in Asian countries, namely India, Japan, Korea, Pakistan, Singapore, and Taiwan. It is apparent that all of these PD programs have similar characteristics and exemplify a critical development in teacher PD in Asia. This development reflects a gradual shift in the center of gravity away from the university-based, supply-side, off-line forms of knowledge production conducted by university researchers for teachers toward emergent school-based, demand-side, on-line, in situ forms of knowledge production by teachers with support from university scholars.

Keywords Teacher professional development · Teachers in Asia

Y.H. Leong (✉) · B. Kaur
National Institute of Education, Singapore, Singapore
e-mail: yewhoong.leong@nie.edu.sg

B. Kaur
e-mail: berinderjeet.kaur@nie.edu.sg

O.N. Kwon
Seoul National University, Seoul, Korea
e-mail: onkwon@snu.ac.kr

B. Kaur et al. (eds.), *Professional Development of Mathematics Teachers*, Mathematics Education – An Asian Perspective,
DOI 10.1007/978-981-10-2598-3_1

1.1 Introduction

"The quality of an education system cannot exceed the quality of its teachers" was one of the three key findings in the McKinsey's report on how the world's best performing school systems come out on top (Barber and Mourshed 2007, p. 16). The report also states that "the main driver of the variation in student learning at school is the quality of teachers" (p. 12). The quality of teachers depends on several factors and one of them is certainly their continuous professional development (PD) (Barber and Mourshed 2007; Mullis et al. 2012).

There has been an interest over the last decades in teacher PD and its impact on both students and teacher learning (Avalos 2011; Yoon et al. 2007). For example, in a most recent study using fourth- and eighth-grade mathematics data from the 2003, 2007, and 2011 TIMSS assessments, Liang et al. (2015) conducted a cross-national empirical study to examine teacher participation in PD and its impact on student achievement. They conclude that professional learning for teachers of fourth- and eighth-grade mathematics is associated with increased student achievement. Based on the positive association between student mathematics achievement and teacher PD, researchers conducted a series of multiple regression models. The regressions indicated that, after controlling for GDP and educational expenditure, there was a statistically significant association in 2007 in five of the six PD areas for fourth-grade students (math content, pedagogy, curriculum, integrating technology, and improving critical thinking and problem-solving skills) and in 2011 in one area. For eighth-grade students, a percentage point increase in access to teachers with PD in mathematics content, pedagogy, curriculum, and integration of technology increased the national mean mathematics achievement score by an average of 1.04, 1.24, 0.93, and 1.07 points, respectively.

Similarly, Unal et al. (2011) analyzed the TIMSS 2007 data of participants from Turkey and found that mathematics teacher PD made a big difference and affected students' performance positively. Kwon and Ju (2012) also claimed that a part of the high performance of Korean students in TIMSS and PISA may be attributed to the higher level of teacher preparation and PD. These findings, when coupled with other studies (for example, Desimone 2009), emphasize the importance of linking the content of professional learning to specific outcomes for students, ensuring depth of teacher content knowledge and content-specific pedagogy, depth of knowledge of curriculum, assessment practices, and technology integration into the content. These studies support the implementation of policies, advocacy, and practices for PD as a vehicle for improving student achievement and supporting educational reform.

The premise of this book is that teachers are the key to students' opportunities to learn mathematics. What mathematics teachers know, care about, and do is a product of their experiences and socialization both prior to and after entering teaching, coupled with the impact of their ongoing professional education. The significance of this impact varies among different education systems: the effects of professional education appear in some systems to be weak or even negligible, whereas other systems are structured to support effective ongoing professional education and instructional improvement. Documenting and focusing on the mathematics teacher PD are important in the context of teachers' central role in students' learning of mathematics. Also important is the fact that efforts to improve students' opportunities to learn mathematics cannot succeed without parallel attention to their teachers' opportunities for learning. Thus, teacher PD is a crucial element in the effort to build an effective system of mathematics education.

The pivotal role of the teacher—and hence teacher education—is attested by a discernible surge in reports on PD in the mathematics education literature over the last decade. Apart from an increase in the number of journal articles and book chapters in this sub-field, there had been special issues of mathematics education journals (e.g., ZDM Special Issue on Evidence-based Continual Professional Development, 2015), scholarly books (e.g., The 15th ICMI Study Group report on the professional education and development of teachers of mathematics, 2009), and even volumes of a handbook (Volumes 1–4 of The International Handbook of Mathematics Teacher Education, 2008) that are focused on this area of PD for mathematics teachers. The collection of chapters in this book further contributes to this trend. In particular, we seek to moderate an underrepresentation in the global corpus on this subject: an Asian perspective.

The outcomes of TIMSS (Mullis et al. 2008, 2012) and PISA (OECD 2010, 2013) show us that students in some Asian countries, such as Korea, Taiwan, Japan, and Singapore, achieve much more than their counterparts in other Asian countries, such as Indonesia and Malaysia. This may be a consequence of varying teacher recruitment standards, teacher preparation and ongoing development programs, and other factors, such as political, social, and cultural factors. As teacher PD does have an impact on student outcomes, this book attempts to provide a resource for scholars to hypothesize relationships between the myriad attributes of mathematics teacher PD and student outcomes. In addition, cross-cultural exchange of knowledge and information about the professional education of teachers of mathematics would be beneficial. Learning about practices and programs in Asian countries can provide important resources for research, theory, practice, and policy in teacher education, both locally and globally.

In the rest of this chapter, we provide a broad overview of the two main parts of the book before concluding with some observations about current trends and likely directions of mathematics teacher PD in Asia.

1.2 Policies, Structures, Frameworks, and Contexts

Part I comprises eight chapters on policies, structures, frameworks, and contexts. These chapters provide us with some ideas about mathematics teacher PD in eight Asian countries: China, Indonesia, Japan, Korea, Malaysia, Philippines, Singapore, and Taiwan. It is apparent from these chapters that there are similarities and also differences in the approaches to mathematics teacher PD in these countries. Table 1.1 shows the various acts and national initiatives that have had an impact on teacher PD, including mathematics teachers, in the respective countries.

Table 1.1 Acts and national initiatives related to teacher PD

Country	Act/Initiative	Year	Mandatory teacher accreditation/PD requirements
China	The Teachers Act	1994	In-service teachers must do a minimum of 240 h of PD over a five year period
Indonesia	Undang-Undang Guru dan Dosen (UUGD) Number 14	2005	The law mandates standard qualifications for teachers to teach in schools and universities
	UUGD Number 14 (Chapter IV Unit 18) Law	2005	All teachers in Indonesia must have a national teaching certificate as a license to practice
			Teachers are free to engage in activities that develop them professionally
Japan	Lesson Study	Early 1900s	A cultural and traditional form of PD that is inherent in the Japanese school system
	Renewal System of Teachers' Certificate	2009	30 h of "certificate renewal courses" approved by Minister of Education, Culture, Sports, Science, and Technology
Korea	The Teachers Act (New Educational Reform Plan)	1995	In-service teachers must receive training at regular intervals. However, the plan did not specify the mandatory hours and intervals
	Reform Act	2010	Development of teaching competence was enacted into law. Teacher evaluation system by students and parents was adopted. Teachers whose student evaluation is 2.5 points or less out of a total of 5 points must undertake 30 h of training
	On-the-job Training	2011	Teachers must complete over 60 h of on-the-job training per year, which is a feature of their performance-based pay
Malaysia	Directive from MOE	2005	Teachers to undergo 7 days of in-service training per year directed by the school leaders
	Malaysia Education Blueprint 2013–2015	2013	All practicing teachers must undergo continuing PD at various stages of their teaching career
Philippines	Department of Education (DepEd) annual in-service (INSET) program	2001	An INSET program ranging 3–5 days per year according to the school calendar issued by the DepEd
Singapore	Thinking Schools Learning Nation (TSLN) vision	1997	As of 1998 teachers are entitled to 100 h of PD per year that is funded by the Ministry of Education
	Teach Less Learn More (TLLM) initiative	2005	Planned time for teachers during curriculum hours to meet, plan, and deliberate on their instructional practices was made mandatory
Taiwan	Education Act	1996	In-service teachers must do 90 h of PD over a period of 5 years
	The Education Act was repealed in 2003	2003	With the dissolution of the act, teachers now participate in PD to improve themselves without any mandatory requirements

From Table 1.1 it is evident that each of the eight countries is at a different phase in its development of teacher professionalism. In some countries there are mandatory acts and regulations governing the continuous teacher PD while in others the situation is lax and in such cases a lot would depend on the teachers' own motivation for development and also the availability of resources.

It is evident from Chap. 2 by Huang, Ye, and Prince that mathematics teacher PD in China has had a long tradition and is structured with very clear development pathways. The Teacher Act of 1994 (Ministry of Education China 1994) has led to the development of an accreditation system which sets guidelines for teachers to obtain their license to practice. There is a ranking and promotion system and teachers have to undertake a minimum of 240 h of PD over a five-year period. The local education authorities stipulate the requirements for teacher continuing education programs according to the ranks of the teachers. The major PD practices are one-to-one mentoring; practice-based research activities comprised of three clusters, namely, routine activity, competitions, and new developments; and both training and education degree programs for teachers to upgrade themselves, implement new curriculum contents and initiatives, and most importantly raise the quality of mathematics instruction in the country. Through PD there is a dedicated attempt to develop expert teachers with deep knowledge in both content and pedagogy.

In Indonesia, as noted in Chap. 3 by Kusumah and Nurhasanah, national certification of teachers was only mandated in 2005. This is probably the first step toward setting standards for teachers to be professionals. The UUGD, Number 14, Chapter IV, Unit 18, states that, "Teachers must have academic qualifications, competencies (pedagogical, social, and professional), national certification for teaching, good physical and spiritual health, and the desired ability to achieve the national education goal" (Depdiknas 2005, p. 6). Although there appear to be no guidelines as to teacher PD, it is evident from this chapter that mathematics teachers do engage in PD mainly through two types of programs. The first are programs that teachers attend outside of their schools that are often courses of study at institutions of higher learning, training workshops and conferences. The two main institutes that provide PD programs for mathematics teachers are the Educational Institute of Quality Assurance and the Institute for Mathematics Teacher Training. The second type allows teachers to engage in learning while carrying on with their duties in school. The authors note that lesson study is a good form of activity for the second type of program. The Japan International Cooperation Agency has been instrumental is initiating lesson study in Bandung, Yogyakarta, and Malang in Indonesia.

Takahashi in Chap. 4 describes lesson study as the fundamental driver for mathematics teacher development in Japan. He outlines the three levels of expertise of mathematics teachers and how lesson study develops teachers at Level 3, which is beyond the scope of any teacher preparation programs in Japan. School-based in-service training is a tradition and culture in Japanese schools, particularly lesson study, during which teachers work collaboratively to develop their pedagogy (Centre for Research on International Cooperation in Educational Development University of Tsukuba [CRICED], n.d.). Although it is not mentioned in this chapter, a recent development that has also fueled the development of teachers is the mandatory

Renewal System of Teachers' Certificate introduced in 2009 (Ministry of Education, Culture, Sports, Science, and Technology Japan [MEXT], n.d.). Every 10 years teachers have to renew their practicing certificates and two years prior to the expiration of the certificate they have to complete 30 h or more of "certificate renewal courses" approved by the Minister of Education, Culture, Sports, Science, and Technology (MEXT, n.d.). The 30 h are comprised of 12 h on core topics of reflection on teachership and understanding how children change; trends in educational policies and coordination and cooperation in and out of school; and 18 h on elective topics related to teaching subjects, guidance to students, and other topics to enrich education (CRICED, n.d.).

From Chap. 5, Kwon, Park, Park, and Park state that teachers in Korea are constantly developing themselves and that the excellent achievement of their students in international benchmark studies such as TIMSS and PISA is a result of teacher development. They must complete at least 90 h of PD activities to upgrade their initial teaching certificate (usually within 3–4 years of their preservice training). Subsequently they are required to participate in PD activities every year (Sami 2013). In 1995, the New Educational Reform Plan was announced that requires that in-service teachers must receive training at regular intervals. However, the plan did not provide any mandatory details about the number of hours and regularity of the intervals, such as every year or three years (Education Reform Commission 1996). In recent years, the government has fully supported a minimum of 20 h of annual PD for each teacher. However, most teachers attend 40–60 h of PD activities to keep abreast of new developments in their fields of expertise.

In 2005, the OECD (2005) reported that teachers in Korea had a low rate of participation in training programs. In response to the announcement of this finding, the government made training programs mandatory. The Ministry of Education and Science Technology (2010) enacted the development of teaching competence into a law. A teacher evaluation system by students and parents was adopted. Teachers whose student evaluation is 2.5 points or less out of a total of 5 points must undertake 30 h of training. In addition, as of 2011, teachers have to complete over 60 h of on-the-job training per year, which is a feature of their performance-based pay.

Chiew and Lim in Chap. 6 state that mathematics teacher PD in Malaysia mainly involves two types of activities. The first type is carried out by education agencies of the Ministry of Education (MOE), which conducts in-service courses and workshops for teachers to accomplish the requirements and changes in the mathematics curriculum. This is meant to ensure that teachers are competent to teach and deliver what is prescribed in the curriculum. The second type is more autonomous and allows teachers to take charge of their own development through research-based projects such as action research and lesson study.

The MOE has begun to try to regulate teacher PD. The directive from the MOE in 2005 mandating seven days of PD per year for teachers was not necessarily subject pedagogy specific as the school leadership was empowered to decide on the focus of the PD and often used it to address the general needs of their schools. Recent developments arising from Malaysia's participation in international benchmark studies such as PISA and TIMSS has led the MOE to examine mathematics instruction in schools. The findings show that there is a lack of an acceptable standard of teaching in schools and thus the MOE (2012) is initiating a reform to transform the landscape of Malaysian education system. This reform, the Malaysian Education Blueprint 2013–2015, is mandating that teachers undergo compulsory continuing PD. As the implementation of the blueprint is still in its infancy, it may be too premature to say how mathematics teacher PD may change.

In the Philippines, as noted by Verzosa, Tulao-Fernando, and Vistro-Yu in Chap. 7, the Department of Education mandates three to five days of PD every year for all teachers. For some teachers this may be the only PD they undertake in the year but may not support in them in their own subjects. Therefore, mathematics teacher PD is often implemented outside of annual school INSET. There appear to be many contexts and opportunities for mathematics teachers to engage in PD but often they lacked long-term coherence.

Kaur and Wong in Chap. 8 recount how two national initiatives of the Ministry of Education in Singapore, the Thinking Schools Learning Nation vision (Goh 1997) and the Teach Less Learn More initiative (Ministry of Education 2005), have fueled teacher PD, including mathematics teacher PD. They outline how the systemic infrastructure put in place by the Ministry of Education has facilitated mathematics teacher PD. Mathematics teachers work and learn collaboratively in the classroom while addressing issues related to the teaching and learning of mathematics and being part of research projects and lesson study groups. They also engage in PD activities to suit their individual needs. They attend higher degree courses at universities in Singapore and elsewhere. They also participate in PD activities conducted regularly by the Association of Mathematics Educators, the Singapore Mathematical Society, and the Academy of Singapore Teachers.

In Taiwan, according to Lin and Chang in Chap. 9, the Education Act of 1996 stipulated that teachers must attend at least 18 h of PD per year or accumulate 90 h over five-years. However, it did not specify any particular PD that teachers must undertake. The act was repealed in 2003 and at present teacher PD is guided by three policy directions: upgrading the academic qualifications of teachers through master degrees, engaging teachers in lifelong learning through workshops and sustained school-based projects addressing the needs of teachers so that they keep abreast of educational issues and reforms and using technology to provide a one-stop resource to support teachers in their professional learning. Since 2003 the decreasing emphasis on the number of hours of PD per year or over a period of five years signals a positive development in teacher PD in Taiwan. Teachers are empowered to take charge of their lifelong learning in the spirit of professionalism.

1.3 Innovative PD Programs in Asia

Part II consists of seven chapters that showcase innovative approaches to mathematics teacher PD in five Asian countries: India, Korea, Pakistan, Singapore, and Taiwan. Table 1.2 shows some of the common characteristics of their approaches.

Table 1.2 Characteristics of PD approaches

	Participants	Site	Model of PD		Key attributes of PD program	
Country	In-service teachers	Teachers' classrooms	Cascade	Hybrid—Training[a] + work in classrooms of teachers	Situated learning	Community of practice
India	√	√	×	√	√	√
Korea	√	√	×	√	√	√
Pakistan	√	√	×	√	√	√
Singapore-1	√	√	×	√	√	√
Singapore-2	√	√	×	√	√	√
Taiwan-1 and 2	√	√	×	√	√	√

[a]Training here refers to sharing or co-construction of knowledge by experts with teachers in the PD program
√ denotes applicable; × denotes not applicable

From Table 1.2 it is apparent that all six PD programs detailed in Chaps. 10–16 (note that Chaps. 15 and 16 are different aspects of the same project in Taiwan) have similar characteristics, such as the participation of in-service teachers, the location of the PD in teachers' classrooms, the PD model being a hybrid where experts share or co-construct knowledge with teachers who integrate the knowledge into their classroom practices either simultaneously or after class, and the key attributes of the PD programs being situated learning and community of practice of teachers. None of the PD programs adopted the "cascade model" (Kumar et al. 2015); instead all of them were of the hybrid model type (Kaur 2011), which draws on the "training model of PD" (Matos et al. 2009) and supports teachers in integrating knowledge from experts or that which has been co-constructed into their classroom practice. This exemplifies a critical development in teacher PD in Asia. This development reflects a gradual shift in the center of gravity away from university-based, supply-side, off-line forms of knowledge production conducted by university researchers for teachers toward emergent school-based, demand-side, on-line, in situ forms of knowledge production by teachers with support from university scholars.

A synopsis of the innovative approaches to PD follows. Kumar and Subramaniam in Chap. 10 present a case study of an in-service teacher who participated in a PD program highlighting the constraints and affordances in bringing about shifts in the teacher's practice toward developing reasoning in mathematics. The PD program consisted of a training portion of workshops conducted by the university scholars, followed by collaboration with the teachers in their classrooms, and finally teachers in the PD program developing fellow teachers in their school who were not in the PD program. The nature of the PD program facilitated learning within the community of practice, which was comprised of teachers, teacher-educators, and researchers. This study shows promise in the potential for developing communities of practice to engage in the enterprise of analyzing and developing teaching of mathematics in schools.

Kwon, Park, Park, and Park in Chap. 11 describe a community-based mathematics teacher PD program that brought together teachers and mentors to work on common goals, thereby developing multitier communities of practice. The three-phase program consisted of a preparatory intensive course and teaching practice and collaboration followed by post-program sharing by participants of their experiences. The PD program was coherent with the needs of the teachers as it supported the present initiatives of the mathematics curriculum in Korea, i.e., STEAM classes and storytelling. In Chap. 12, Halai elaborates on an innovative PD program that draws on the classroom as a site for teacher learning in Pakistan that sparked the emergence of a paradigm shift in mathematics teacher education in Pakistan. The PD was an advanced diploma program for mathematics teachers that consisted of both training and practice aspects of development. From this chapter it is evident that in-service education and continuing PD with a strong component of a school-based practicum offers a way forward for teacher education that is absolute for teachers and schools.

Kaur, Bhardwaj, and Wong in Chap. 13 outline in great detail their teaching for metacognition-themed PD program in Singapore. Based on the hybrid model of PD, the program has three phases with teachers (classroom, lead, and master), curriculum specialists, a researcher, and a professor working together in two-tier communities of practice. The three phases of the project are training (acquisition of knowledge), integration of knowledge into classroom practice, and empowerment of teachers to develop fellow teachers. The last phase is critical for teachers to sustain their PD and induct others into practices that lead to worthy student outcomes (Kaur 2015). In Chap. 14, Leong, et al. note that for instructional innovations to take root in mathematics classrooms, curriculum redesign and teacher PD are two necessary and mutually reinforcing processes: A redesigned curriculum needs to be seen as an improvement in order to facilitate teacher buy-in—an ingredient for effective PD; on the other hand, teacher PD content needs to be directed toward actual usable classroom implements through the enterprise of collaborative curriculum redesign. In their chapter, they examine the interaction between researchers and teachers in this collaborative enterprise through the metaphor of boundary crossing. In particular, they study a basic model of how "boundary objects" located within a "replacement unit" strategy interact to advance the goals of PD.

Chapter 15 by Lin, Hsu, and Chen and Chap. 16 by Chen and Lin both are based on the same nation-wide PD program in Taiwan. The Lighten-up School Based Program (LUSBP) adopts a design-based approach and consists of tiers of educators, teachers, and students who collaborate as communities of practice and engage in crafting tasks, enacting them, and reflecting on them using student work as an input and revising them for subsequent work in mathematics classrooms. The findings of the project hold promise for school-based PD, as it facilitates teacher growth and also the development of teacher-educators who experience the integration of their expert knowledge through the teachers in the classrooms and the tensions that may arise during the process. Chen and Lin in Chap. 16 describes how two schools participating in the LUSBP worked with diagnostic conjecturing activities during their PD. From the findings of the study of the two schools it is apparent that the teachers' learning from the workshops designed to equip them with knowledge about conjecturing activity know-how was integrated into the classroom practices of the teachers. As teachers worked in their respective communities of practice they supported and also challenged each other to develop and refine their student-centered teaching practice.

1.4 Interacting but Diverse Asia

From the broad review in the above sections, it appears that PD practices within some jurisdictions in Asia have reached a rather stable state. The example of teacher PD in Japan is a case in point. In almost all reports about teacher PD in Japan, including the one included in this volume by Takahashi (Chap. 4), there is a unified and relatively unchanging image of how PD among mathematics teachers are conducted: Lesson Study. According to a number of Japanese writers, the origin of Lesson Study is traceable to the year 1872. Against the global norms of educational shifts—and hence the shifting forms of PD practices—the relative stability of Lesson Study across temporal and geographical zones within the country stands as a rather unique model of sustainable teacher PD.

There is evidence that, since about two decades ago, Lesson Study as a form of teacher PD has been 'imported' to other Asian countries (e.g., Isoda et al. 2007). Further evidence can be found in this volume—Chap. 3 by Kusumah and Nurhasanah on mathematics PD Programmes in Indonesia, Chap. 6 by Chiew and Lim on mathematics teacher PD in Malaysia, and Chap. 8 by Kaur and Wong on PD of mathematics teachers in Singapore.

However, other chapters that describe PD strategies at national level did not mention influences from the enterprise of Lesson Study. These include South Korea (Chap. 5), the Philippines (Chap. 7), and Taiwan (Chap. 9). Moreover, there were also no reference to Lesson Study in the chapters that detailed specific PD implementations in India and Pakistan (Chaps. 10 and 13, respectively). Even in the case of Singapore where Lesson Study was ostensibly foregrounded, it was described as one of a number of other platforms (such as Learning Circles and Action Research) for ongoing school-based PD. In addition, other emerging forms of PD in Singapore such as the "Replacement Unit Strategy" were also being developed (Chap. 15).

In other words, while there is interaction across the continent in terms of exchanging ideas on PD strategies, countries within Asia are embarking on their respective journeys with regards to finding their own way in mathematics teacher PD. Given their different historical–political trajectories, it is not surprising to read about diverse practices of teacher PD in different regions of Asia. Thus, while perhaps the conditions in Japan would conduce to uniformity and continuity in PD practices, in most other parts of Asia, there are more experimentation and openness in developing PD strategies that suit the sociocultural contexts of the respective jurisdictions.

Fundamental challenges also vary widely across the continent. In some regions, the effectiveness of PD at scale is threatened by macro-level issues such as national education policy reversals and population-wide low commitment to high-standard education. In other jurisdictions, the attention in PD efforts are directed at more 'local' challenges such as the networking of teachers into learning communities and the tailoring of PD programmes to render them more relevant to the actual instructional practices of the teachers.

1.5 Convergences in Asia

Despite the diversity, the chapters in this volume point to some convergences of mathematics teacher PD in the region. First, almost all the PD programmes reported in this book included components which were school-based—and which were designed to address issues that resonated with teachers' experiences in practice. This aspect of nearness-to-practice is inherent in Lesson Study; thus, the countries that utilised this enterprise as a prominent platform for PD—such as Japan, Indonesia, Malaysia, and Singapore—are included in this class; but it is certainly not limited to Lesson Study. For example, projects in China and Korea involved "practice-based" or "practice-oriented" approaches to teacher PD. In China, PD for teachers—at all levels of expertise—was conducted primarily in a context where teachers observe and discuss about exemplary teaching within classroom instructional situations; the Korean project (as described in Chap. 11) focused on an instructional innovation of "story telling" by examining its use and modification in actual classroom use by teachers.

The Lighten-Up School-Based Program (LUSBP) in Taiwan is another example of a project that emphasized school-based approaches in PD. Teachers participated in PD in this programme through the process of planning instructional materials, implementing the materials in their classes, and reflecting upon the implementation for learning and changes for subsequent teaching. In the case of the project located in Pakistan (Chap. 12), the PD for practising teachers included a "Practicum" component. This is novel in that we would usually associate Practicum with pre-service teacher education. Through interactions between what was taught during PD classroom settings and Practicum teaching, teachers in the project reflected on how newly introduced ideas can be incorporated into their classroom practice.

That this movement toward school-based approaches in PD practices within some parts of Asia is a rather recent phenomenon is acknowledged by the authors of a number of chapters in this volume (e.g., Chaps. 12, 13 and 16). This signals a likely significant shift from traditional conceptions of PD being university-based course offerings to one where actual problems of practice become the objects of inquiry in teacher learning.

Closely related to this trend toward nearness-to-school is the "collaboration" between mathematics teacher-educators (usually working in university contexts) and mathematics teachers (in schools) in redesigning mathematics curriculum or instruction. Collaboration in this sense is conspicuous in a majority of the reports found in this volume. Teacher PD is located in such collaborations as teachers interact with other experts outside of the school system. Quite different from the model of teachers as passive receivers of knowledge—associated with traditional in-service courses given by university professors, this emerging model, as represented by the "collaboration" metaphor, encourages teacher learning by working with these experts—and in the process, they examine new knowledge domains, reflect on existing teaching practices, and clarify novel teaching approaches.

1.6 Distinctively Asia?

It may be argued that the "convergences" mentioned in the above section are not distinctly an Asian phenomenon, but really a global movement. All over the world, PD practices are tending toward nearness-to-school and collaboration with teachers (e.g., Robutti et al. 2016; Weißenrieder et al. 2015). But when we examine closer the kind of school-based and collaborative work that researchers do—as reported in the chapters here—we find clues for points of departure at various areas. First, the PD programmes were usually conducted within the broader framework of a national or provincial vision of quality pedagogical practices. This means that the researchers entered the school with a prior interpretation of how this vision can be translated into actual instructional practices in the classroom. In other words, the researchers began the engagement with schools with a rather clear portrait of how the schools can realise the educational ideals of policy makers. Second, and closely related to the first, "collaboration" with teachers based on this model of engagement means that the researchers proposed and charted the agenda of instructional design, while the teachers provided inputs for tweaking some aspects of the design. This picture of collaboration is not one of complete equality of roles and voices. It is a model where university mathematics educators were regarded as possessing knowledge—of both mathematics content and pedagogical expertise—at a higher vantage point, teachers usually learnt from these experts and less the other way round, and teachers contributed to the enterprise by highlighting practical constraints, and sharing their learning experiences.

We think a depiction of this rather distinctively Asian way of doing school-based and collaborative PD is "pragmatic". We use this term in two senses: (1) Researchers are less locked-into traditional paradigms of research, neither is research in teacher education oriented toward formulation or development of global theories (as in abstractions for the purpose of universal applicability). Rather, teacher PD work proceeds along the lines of pedagogical fundamentals that tapped on a range of disciplinary traditions. Thus, the use of eclectic frameworks is the norm; (2) the whole enterprise of teacher PD is goal-driven in deference to policy objectives: the main goal is to improve instructional practices in the classroom—toward policy ideals—through teacher PD. What counts as quality instructional practices are centrally crafted and usually resides in the higher reaches of the policymaking structure. During PD, these parameters are interpreted into instructional designs that are sensitive to teachers' authentic practices.

This pragmatic approach is perhaps most appropriate for societies—such as in most Asian countries—with a history of more centralized administration. Within this model, mathematics teacher educators tap on a rather stable and widely-accepted repertoire of theoretical starting points—as supported by the administration, and focus on teacher PD that integrates these ideals into the realistic framework of teachers' practices. This approach channels less resources on negotiation between researchers and teachers and more resources on making 'it' work in the classroom.

At this stage, it is unclear if this pragmatic approach—arguably the Asian way—would make way for other ideologically based methods of teacher PD. This next step of the Asian journey in teacher PD may contribute to the larger rhetoric of whether 'east' and 'west' would find their respective paths of development or they would ultimately converge.

References

Avalos, B. (2011). Teacher professional development in "teaching and teacher education" over ten years. *Teacher and Teacher Education, 27*, 10–20.

Barber, M. & Mourshed, M. (2007). *How the world's best performing school systems come out on top*. McKinsey & Company.

Centre for Research on International Cooperation in Educational Development University of Tsukuba (CRICED). (n.d.). *Education system and practice in Japan*. Retrieved from http://www.criced.tsukuba.ac.jp/keiei/kyozai_ppe_f8html

Depdiknas (Department of National Education). (2005). *Undang-Undang Republik Indonesia No.14 tahun 2005* (Law of Republic Indonesia about Teachers and Lecturers). Jakarta: Departemen Pendidikan Nasional.

Desimone, L. M. (2009). Improving impact studies of teachers' professional development: Toward better conceptualizations and measures. *Educational Researcher, 38*(3), 181–199.

Education Reform Commission. (1996). *Report on new education reform*, Education Reform Commission.

Goh, C. T. (1997). Shaping our future: "thinking schools" and a "learning nation". *Speeches, 21* (3), 12–20. Singapore: Ministry of Information and the Arts.

Isoda, M., Stephens, M., Ohara, Y., & Miyakawa, T. (2007). *Japanese lesson study in mathematics: Its impact, diversity and potential for educational improvement*. Singapore: World Scientific.

Kaur, B. (2011). Enhancing the pedagogy of mathematics teachers (EPMT) project: A hybrid model of professional development. *ZDM—The International Journal on Mathematics Education, 43*(7), 791–803.

Kaur, B. (2015). What matters? From a small scale to a school-wide intervention. *ZDM—Mathematics Education, 47*(1), 105–116.

Kumar, R. S., Subramaniam, K., & Naik, S. (2015). Professional development workshops for in-service mathematics teachers in India. In B. Sriraman, J. Cai, K. H. Lee, L. Fan, Y. Shimizu, C. S. Lim, & K. Subramaniam (Eds.), *The first sourcebook on Asian research in mathematics education: China, Korea, Singapore, Japan, Malaysia and India* (Vol. 2, pp. 1631–1654). North Carolina: Info Age Publishers, Charlotte.

Kwon, O. N., & Ju, M. K. (2012). Standards for professionalization of mathematics teachers: Policy, curricula, and national teacher employment test in Korea. ZDM. *The International Journal on Mathematics Education, 44*(2), 211–222.
Liang, G., Zhang, Y., Huang, H., Shi, S., & Qiao, Z. (2015). Professional development and student achievement: International evidence from the TIMSS data. *Journal of Postdoctoral Research, 3*(2), 17–31.
Matos, J. F., Powell, A., & Sztajn, P. (2009). Mathematics teachers' professional development: Processes of learning in and from practice. In R. Even & D. L. Ball (Eds.), *The professional education and development of teachers of mathematics* (pp. 167–183). New York: Springer.
Ministry of Education. (2005). *Teach less learn more*. Singapore: Author.
Ministry of Education of China (1994). *The teachers ACT in P. R. China*. Retrieved December 12, 2015 http://www.moe.edu.cn/s78/A02/zfs__left/s5911/moe_619/tnull_1314.html. (in Chinese)
Ministry of Education, Culture, Sports, Science, and Technology Japan (MEXT). (n.d.) *Improving the quality and ability of teachers*. Retrieved from http://www.mext.go.jp/english/elsec/_icsFiles/afieldfile/2015/08/11/1303528_01_pdf
Ministry of Education and Science Technology. (2010). *2010 Development of teaching competence standard manual*. Seoul: Ministry of Education and Science Technology.
Ministry of Education Malaysia (2012). *Preliminary Report Malaysia Education Blueprint 2013–2025*. Ministry of Education Malaysia. Retrieved on December 12, 2015 from http://www.moe.gov.my/userfiles/file/PPP/Preliminary-Blueprint-Eng.pdf
Mullis, I. V. S., Martin, M. O., & Foy, P. (2008). *TIMSS 2007: International mathematics report*. Chestnut Hill, MA: TIMSS & PIRLS International Study Centre, Boston College.
Mullis, I. V. S., Martin, M. O., Foy, P., & Arora, A. (2012). *TIMSS 2011: International mathematics report*. Chestnut Hill, MA: TIMSS & PIRLS International Study Centre, Boston College.
OECD. (2005). *Teachers matter*. Paris: OECD.
OECD. (2010). *PISA 2009 Results: What students know and can do: Student performance in reading, mathematics and science* (vol. 1). OECD Publishing.
OECD. (2013). *PISA 2012 Results: What students know and can do: Student performance in mathematics, reading and science* (vol. 1). OECD Publishing.
Robutti, O., Cusi, A., Clark-Wilson, A., Jaworski, B., Chapman, O., Esteley, C., et al. (2016). ICME International survey on teachers working and learning through collaboration. *ZDM—The International Journal on Mathematics Education, 48*(5), 1–69.
Sami, F. (2013). South Korea: A success story in mathematics education. *MathAMATYC Educator, 4*(2), 22–28.
Unal, H., Demir, I., & Kilic, S. (2011). Teachers' professional development and students' mathematics performance: Findings from TIMSS 2007. *Procedia Social and Behavioral Sciences, 15*, 3252–3257.
Weißenrieder, J., Roesken-Winter, B., Schueler, S., Binner, E., & Blömeke, S. (2015). Scaling CPD through professional learning communities: Development of teachers' self-efficacy in relation to collaboration. *ZDM—The International Journal on Mathematics Education, 47*(1), 27–38.
Yoon, K. S., Duncan, T., Lee, S. W-Y., Scarloss, B., & Shapley, K. L. (2007). *Reviewing the evidence of how teacher professional development affects student achievement*. Washington, DC: U.S. Department of Education, Institute of Education Sciences, National Center for Education Evaluation and Regional Assistance, Regional Educational Laboratory Southwest.

Part I
Policies, Structures, Frameworks and Contexts

Chapter 2
Professional Development of Secondary Mathematics Teachers in Mainland China

Rongjin Huang, Lijun Ye and Kyle Prince

Abstract This chapter examines major practices of the professional development (PD) of mathematics teachers and relevant supporting infrastructures in Mainland China. The first section provides an overview of the teacher education system in China including teacher preparation, teacher licensure, ranking system, and continuing education. The second section focuses on the main practices of PD programs and their latest developments, which include apprenticeships, hierarchical teaching research, lesson studies and lesson contests, and master teacher development programs. Finally, the authors synthesize a model depicting the system of PD in Mainland China, and discuss what could be learned from Chinese practices.

Keywords Teacher professional development of mathematics teachers · Mainland China · Professional ranking system · Teaching research system · Apprenticeship · Chinese lesson study · Master teacher development

2.1 Background

The efforts to improve student learning in mathematics led researchers to investigate high-achieving education systems and practices in East Asia, including China (Bednarz et al. 2011; Leung and Li 2010; Mullis et al. 2012; OECD 2010). In particular, studies have focused on how Chinese learn mathematics (Fan et al. 2004) and teach mathematics (Li and Huang 2012). However, less attention has been given to how Chinese prepare and develop mathematics teachers (Huang et al.

R. Huang (✉) · K. Prince
Middle Tennessee State University, Murfreesboro, USA
e-mail: rhuang@mtsu.edu

K. Prince
e-mail: kmp3f@mtmail.mtsu.edu

L. Ye
Hangzhou Normal University, Hangzhou, China
e-mail: yeatsylj@126.com

B. Kaur et al. (eds.), *Professional Development of Mathematics Teachers*, Mathematics Education – An Asian Perspective,
DOI 10.1007/978-981-10-2598-3_2

2010; Liang et al. 2013), which is a crucial factor influencing students' mathematics learning and achievement (National Mathematics Advisory Panel [NMAP] 2008; Schmidt et al. 2008; Sullivan and Wood 2008). Although Ma (1999) found that Chinese elementary mathematics teachers have profound mathematics knowledge and skills needed for teaching, recent studies suggest that Chinese mathematics teachers do not have adequate preparation in their pedagogical content knowledge (Li et al. 2008). One possibility is that Chinese practicing teachers develop their mathematics knowledge for teaching during practice (Li and Huang 2008).

Studies have revealed some unique features of the mathematics teacher professional system (PD) in China (Huang et al. 2010; Stewart 2006) such as a well-established ranking and promotion system (Li et al. 2011a), institutionalized teaching research system (Yang and Ricks 2012), and ubiquitous public lesson development (Huang et al. 2011; Liang 2011). This chapter aims to provide a comprehensive and updated picture about practices of PD in Mainland China. First, we provide an overview of the teacher education system in China as a background for understanding practices of PD. Then, we examine major practices of PD of mathematics teachers and their latest developments. Finally, we synthesize a model of the PD system and discuss what other education systems may learn from Chinese practices.

2.2 Overview of Teacher Education System in China

This introduction of teacher education system in China includes degree requirements, teacher preparation programs, accreditation system, and ranking system.

2.2.1 Education Degree Requirements

The Teacher Act (Ministry of Education [MOE] 1994) defines teachers' positions, qualifications, and responsibilities. The *Regulations of Teachers' Qualification* (MOE 1995) further states the teacher professional ranking system, and promotion criteria and procedures. According to *The Teacher Act*, the minimum education required for elementary teachers is completion of a three-year program offered by normal schools; the minimum for middle school teachers is completion of a three-year program offered by normal colleges; the minimum for high school teachers is a four-year bachelor's degree. However, with the development of economics and technology, requirements have increased. Primary school teachers are trained in three-year teacher colleges or four-year universities; secondary (including middle and high) teachers are trained in four-year universities, while some high school teachers are even required to attain a postgraduate degree (Wang 2009). Some developed regions in China aimed to implement "one level of institute (four-year university), two levels of degree (bachelor and master degree for school teachers at all levels).” Typically, primary school mathematics teachers are trained in a college of education at a three-year college or four-year university; secondary mathematics teachers are trained in mathematics departments at four-year universities. The Ministry of Education (1999a, b, 2000) has encouraged comprehensive universities to create teacher preparation programs. Since then, an increasing number of teacher candidates have graduated from comprehensive universities.

Secondary mathematics education programs emphasize providing the preservice teacher with a profound mathematics knowledge foundation and highly advanced mathematics literacy and reviewing and studying of primary mathematics but provide a limited teaching practicum program (Li et al. 2008; Liang et al. 2013). The practice of emphasizing mathematics content knowledge with less attention to pedagogical knowledge and teaching practicum may reflect the beliefs that “university is the best place for learning advanced mathematics content, the core purpose of teacher preparation is to learn subject knowledge because prospective teachers can develop their pedagogical knowledge from their future teaching practice” (Wang 2009, p. 181).

2.2.2 *Accreditation System*

To obtain a teaching license after earning a bachelor degree (in any areas), aspiring teachers are required to pass both written and oral exams (MOE 2011, 2013a). The written exams include the following content components: Synthesis quality, educational knowledge and skills, subject knowledge, and instructional abilities [mathematics knowledge (41 %), curriculum knowledge (18 %), mathematics teaching knowledge (8 %), and mathematics teaching skills (33 %)].

Once passing the written exams, a teacher candidate is eligible to participate in an interview that further examines knowledge and skills such as professional ethics, dispositions, manners, communication skills, and teaching skills. The interview is composed of two parts. Part A focuses on one structured interview question and one topic for mini teaching while part B takes the form of question and answer with an expert panel (From 2011 to 2013, average pass rates are 35 % for the written exams and 70.9 % for the oral exam) (Wu and Ge 2015). A teacher candidate who passes the written exams and the oral interview obtains a teaching license that will need to be revalidated and registered every 5 years (MOE 2013b).

2.2.3 *Ranking and Promotion System*

Similar to university faculty promotion systems, professional promotion systems for primary and secondary teachers have been practiced for decades (MOE 1995). According to the regulations (MOE 1995), there are different professional ranks for secondary and primary teachers. For example, the positions within secondary teacher include senior-rank teacher, secondary level 1, and secondary level 2. For each, there are specifications with respect to political, moral, and academic standings. Local educational authorities provide detailed and specific requirements for promotion at each level. Apart from the described three levels of position rankings, there are some honorary rankings such as "exceptional teacher." They are the model and experts in terms of morals, educating students, and teaching expertise (see Li et al. 2011a, b for details). As part of the promotion system, teachers have to participate in at least 240 h of professional development over a five-year period (MOE 1999a). Local education authorities determine requirements for teacher continuing education programs for different rank teachers.

In August 2015, the Ministry of Human Resources and Social Security [MHRSS] and Ministry of Education [MOE] of China (MHRSS and MOE 2015a) jointly rereleased a document, *The Guidance for Deepening the Reform of Secondary and Primary School Teachers' Promotion System* which unifies the ranks for both secondary and primary teachers into three levels that are aligned with other professional ranking systems. The senior-rank level includes full-senior and senior teachers. The intermediate-rank level is called level 1 teacher, and the primary-rank level consists of level 2 and level 3 teachers. For example, full-senior teachers should meet the following criteria: (1) have high, professional aspirations and firm professional beliefs; have experience working as a teacher for a long time and serve as a guide and steering role in prompting students' growth, and have been an excellent class supervisor and student counselor, and made a great accomplishment in educating students; (2) have a profound understanding and mastering of curriculum standards and subject knowledge; achieved excellent performance in education and teaching, demonstrated an adept in teaching arts, and developed a unique teaching style; (3) have an ability to organize and guide education and teaching research; achieved creative results in educational ideas, curriculum reform, teaching methods, and applied them in teaching practices, and exerted a demonstration and steering role; (4) make exceptional contributions to mentoring and cultivating teachers at level 1, 2, and 3; maintain a high reputation in subject teaching, and have been well-recognized as an education and teaching expert; and (5) normally hold a bachelor or above degree, and have served as an advanced teacher at least five years. For another example, a level 3 teacher (entry level) should meet the following criteria: (1) basic mastery of the principles and methods of educating students, and should be able to appropriately educate and guide students; (2) have educational, psychological, and pedagogical knowledge, and basic mastery of subject matter knowledge and pedagogical knowledge in the subject being taught, and be able to teach a subject; and (3) hold an associate degree or above, and one year successful teaching probation.

Building on the previous regulations (MOE 1995), the recent guidance (MHRSS and MOE 2015a) stresses teacher morality, practical accomplishment, and practical experience, and de-emphasizes academic articles and academic degrees. It suggests the establishment of an evaluation system based on peer expert panel's evaluation. The panel should include highly respected education and instruction experts and experienced teachers. The evaluation forms should include explaining and delivering lessons, interviews, and expert reviews. The local authorities' implementation plans should be approved by MHRSS and MOE by the end of 2015. The transfer to, and first implementation of the new system will take place in 2016. And the new system will be fully implemented after a two-year exploration (MHRSS and MOE 2015b).

2.3 Major Practices of Professional Development System

The well-structured and hierarchal teacher professional development system includes: induction programs, practice-based teaching research programs, upgrading degree program, new curriculum training programs, and master teacher training programs.

2.3.1 *One-to-One Mentoring Practice*

A new teacher is usually assigned an experienced teacher as a mentor for the first couple of years (2–3 years). The idea underlying the *one-to-one mentoring practice* is that novice teachers can learn from experienced teachers regarding routine of teaching, basic skills of lesson design, implementation, evaluation, and understanding of textbooks. At the beginning of each academic year, schools form pairs of mentor-mentees and have a celebration for them. Both mentors and mentees sign formal pacts, which describe the responsibilities of both (Huang et al. 2010). The practice not only familiarizes novice teachers with teaching routines and develops their basic teaching knowledge and skills, but it also exposes experienced teachers to innovative teaching ideas and new instructional technologies (Huang and Li 2008; Wang and Paine 2006).

2.3.2 *Practice-Based Teaching Research Activity*

Teaching research activities refer to various activities of professional development institutionalized by four hierarchical organizations: province/city, district/county, school, and lesson plan group. These organizations (*Jiaoyan Shi*) are responsible for guiding teaching research activities, overseeing teaching administration in schools on behalf of educational bureaus, providing consultation for educational authorities, mentoring the implementation and revision of new curricula, building the bridge between modern educational theories and teaching experiences, and promoting high-quality classroom instruction (Huang et al. 2010; Wang 2009; Yang and Ricks 2012). School-based teaching research groups (TRG) are the basic organizations that organize teaching research activities. We describe teaching research activities in three clusters: routine activity, competitions, and new developments.

2.3.2.1 Routine Teaching Research Activity

There are school-based and across school teaching activities. The school-based teaching activities include two forms: mathematics teaching research groups and grader-based (with other subjects) lesson plan preparation groups. Mathematics teaching research groups are responsible for designing and implementing the teaching and research schedule in each semester or academic year. Grade-based lesson plan groups are responsible for the organization and guidance of lesson plans at that grade. There is a variety of teaching research activities beyond school level (city, district/country levels). For example, in a big southeast city, an academic year includes four big activities in a semester, one activity per month (These activities usually were scheduled on Friday morning). In each activity, the theme and facilitator are preassigned so that each activity can focus on a specific topic. Knowledgeable experts (master teachers, teaching research educators, and faculty from universities) are invited to give lectures or model lessons for teachers. These activities often focus on analyzing textbooks and teaching reference materials, and exploring effective strategies of teaching with an emphasis on college entrance exam preparation (Huang et al. 2010).

2.3.2.2 Various Competitions Focusing on Teaching Skills

Besides routine teaching research activities, educational authorities, and teaching research institutes organize various competitions focusing on teaching skills at both local and national levels. They include lesson competitions, explanation lessons, and explaining problem solving.

Teaching competition classes held for young teachers (under 30 years old) at school, district, city, province, and national levels. Teachers who are under 30 years old are encouraged to participate in teaching competitions. Teachers who participate in the competition select a topic from a predetermined list. Winners at the school level are recommended to participate in the district level competition; winners of district level competitions move up to the city level competition; winners of city level competitions proceed to the province level of competition; finally, the winners of the province level competitions go to compete in the national level competition. However, the competition lessons are collaborative results. For instance, a competition lesson taught at the city level is a collaborative effort of master teachers and teaching researchers within a district. Liang (2011) concluded that these activities provide teachers with opportunities to share their ideas, constantly conduct reflective thinking on teaching, and learn actively from each other. By examining a nationwide lesson competition, Li and Li (2009) found that that lesson competitions could improve teaching to better align with the criteria of an exemplary lesson valued in Chinese culture.

The explaining lesson competition is a variation of lesson competition in which the participant teachers analyze teaching content, identify teaching objectives, select appropriate teaching methods, explain the major processes of teaching, and explain assessment of student learning during the lesson. Similar to lesson competition, there are multiple levels of explaining lesson competition (district, city, province and nationwide). Peng (2007) concluded that lesson explaining could promote participants' growth in subject knowledge as well as pedagogical content knowledge.

Recently, explaining problem solving competitions have developed into a popular way of promoting young mathematics teachers' growth. The participant teachers are required to explain a set of interconnected problems with regard to: the background of problems including context, knowledge, and thinking methods; making sense of the problems; explaining methods of solving the problems, extending the problem by varying conditions, results or analogizing thinking methods, and reflecting on critical steps of problem solving and student difficulties (Ye et al. 2011).

2.3.2.3 New Developments

Recently, teaching research activities have evolved into a new phase, school-based study, by adopting the concept of "communities of practice" (Wenger 1998). The school-based study has shifted its foci (Gu 2005): (a) from development of teaching procedural proficiency to building teaching culture and community, (b) from studying of content and pedagogy to studying of students' learning and teachers' behaviors, (c) from perfecting teaching activities to the cultivation of teaching research awareness, (d) from the reflection on limited experience to upgrading education ideas and cultural reconstruction. In school-based study, practicing teachers are expected to experience the entire process of practical problem posing and solving and develop their practical wisdom and theoretical knowledge. Thus, a safe, trustful, and constructive practical community is emphasized so that practicing teachers and experts can freely share their options about improving teaching and student learning rather than critiquing teachers' behaviors.

The development of open lessons includes the cyclical process of collaborative lesson planning, trial teaching, post-lesson reflection, and revision. In "*action education*" (Gu and Wang 2003), a new approach was recommended: teachers are asked to plan a lesson in his/her own way and teach it to find the gap between the vision of teaching and learning in the new curriculum reform, and his/her own actual teaching and student learning. Based on identifying the teacher's gap, the teacher works with the members of TRG to redesign the lesson in alignment with the new curriculum and theories of teaching and learning. The second lesson is intended to bring about improved teaching and help teachers collect evidence of the gap between theoretical visions and actual teaching and student learning. A third lesson is taught so that the teacher can apply the improved practice to focus on the evidence of student learning and its result (Huang and Bao 2006). This is often coined as Chinese lesson study. In addition to similarities shared with the Japanese lesson study in terms of their activity structures, this approach focuses "on both content and pedagogical knowledge and skills, and an open, learner-centered implementation component" (Lerman and Zehetmeier 2008, p. 139).

Parallel Lesson Study (PLS) is an enriched Chinese Lesson Study. Through the process of Chinese Lesson Study, at least two lesson study groups independently develop exemplary lessons of teaching a selected topic. Then, a teaching research activity at the cross-district level is organized and participants in lesson study groups and others are invited to participate in the teaching research activity. A post-lesson meeting focuses on comparing and contrasting the exemplary lessons. Huang et al. (2014) found that practicing teachers could develop their competence in identifying instructional objectives, improving instructional process, selecting and sequencing mathematical tasks, and developing professional vision through participating in a parallel lesson study.

2.3.3 Implementing New Curriculum Training Programs

China has launched its new *Mathematics Curriculum Standard for Compulsory Education Stage* (MOE 2001a) in 2001, and has officially implemented it nationwide since 2010. There are various training programs helping teachers to understand the reform-oriented curriculum and textbooks, and develop their knowledge and skills for implementing the new textbooks. It is required that all teachers attend at least 40 h of training before using the new textbook (MOE 2004). A "cascade model" has been adopted to train teachers for using the reform-oriented curriculum. It started with "seeding" the new idea and strategies of implementing reform-oriented curriculum and textbooks through the training of "trainers and key teachers" at national level, and then training of "local trainers" through national "trainers," and then the local trainers facilitate the training of classroom teachers (see Huang et al. 2010 for details).

2.3.4 Upgrading Education Degree Programs

There are different channels for unqualified teachers to advance their degrees. Some continuing education colleges at universities have provided programs to attain education degrees through distance education since 2003. In addition to helping teachers upgrade to a required education degree, some master programs or masters in education programs specializing in mathematics education, have been opened in many provinces since 1998. The candidates who complete master programs will get a certificate and can apply for a master's degree if they pass an English test and thesis defense (Huang et al. 2010).

2.3.5 *Training Expert Teacher Programs*

Expert teachers refer to those who have an advanced professional title or above, and have played key roles in teaching research activities in their school, district, or city. Expert teachers include *backbone teachers* and *master teachers*. Backbone teachers refer to those who are young with an advanced rank, and are actively involved in various teaching research activities. Master teachers conventionally refer to those who are of higher reputation and play an influential role in a certain region. They have modern education and instructional notions, super practical ability, hold a professional title of exceptional teaching, are educational experts at city level, middle-age subject leaders, or have made exceptional contributions and enjoyed governmental subsidy (Quan 2009).

2.3.5.1 Backbone Teacher Training Program

Ministry of Education launched *backbone teachers* programs (MOE 1999a) to advance the quality of education. The Ministry of Education has taken the responsibility for training 10,000 *backbone teachers* nationwide. Local governments have also made efforts to train *backbone teachers* through raising specific funds. At the national level, this program was organized and implemented by several key normal universities. The courses for backbone teacher training programs include theory and skill (40 %), practice and observation (30 %), education research (30 %) (MOE 2001b). At provincial and municipal levels, the key teacher program was organized in a similar way by local universities or institutes.

2.3.5.2 Master Teacher Training Program

After more than 10 years of training backbone teachers, there are a great number of backbone teachers who have taken leadership in their schools. Some of them have developed as master teachers. To make use of the existing master teachers, and develop more master teachers, training programs have emerged and become popular. These programs are either organized by university and school partner (*master teacher training program*), or are led by a master teacher *master teacher workstation* (*MTW*). Master teacher workstations consist of a master teacher (recognized and conferred by Provincial education authority) and several backbone teachers (recommended by local schools).

Although there is great variation in master teacher workstations, there are three essential commonalities (Quan 2009): the master teacher leads the practical community, all members of the community share common goals, and members have an identity of ownership in the community. The major activities in a MTW include: (1) master-led activities (master demonstrates teaching, master presentation, master comments on lessons), (2) collaborative teaching research (evaluation of classroom teaching, examining student and teacher learning through surveys and interviews; classroom observation across subjects; using online platform to share and collaborate); (3) reading and reflection (reading professional articles and books, and reflecting and sharing what they read, teach, and observe), and (4) project-driven activities (differentiated development plan, exploring specific tasks or projects) (Quan 2009). Li et al. (2011b) examined a MTW focusing on elementary mathematics teachers at the province level, and found that the MTW can help participating teachers develop a deeper understanding of mathematics and pedagogical content, and mathematics and its structure in textbooks. Huang and Huang (in press) described how a MTW helped an experienced teacher to develop a problem posing-based teaching approach from initial experience, via, intended exploration of effective strategies in reviews, and continued exploration of the teaching approach in mathematic teaching in general.

Some universities initiate (work with master teachers at practicum partner schools) master teacher training programs. Take a high school mathematics master teacher training program organized by a Normal University in southeastern China for example. The training program consists of five models: (1) theory learning (theories of mathematics learning; reform-oriented models of mathematics teaching, mathematics curriculum), (2) practical instruction, (3) study abroad; (4) field investigation in other provinces; (5) demonstrating teaching; (6) research projects; (7) demonstrating teaching in rural areas; (8) accomplishment demonstration (1200 h in total).

2.4 Conclusion and Discussion

2.4.1 *Summary*

The systematic ranking promotion system and hierarchical teaching research system are two fundamental infrastructures for supporting teachers' professional development. Li et al. (2011a, b) explained why this ranking system works in China from political, academic, and cultural perspectives. From a political perspective, this system provides a bottom-up, manageable mechanism for teachers to pursue higher professional ranks with an increase of benefits. From an academic perspective, the ranking system specifies what professional competences are needed at each rank so that teachers know what they need to improve in order to get a promotion. For a cultural perspective, teachers get used to getting promotions through examinations and being analyzed in public. Teachers respect senior or master teachers and are willing to learn from others, particularly knowledgeable master teachers.

We further synthesize the trajectory of teachers' growth as follows (Fig. 2.1): Novice teachers typically possess strong subject knowledge through teacher preparation programs. With the support of one-to-one mentoring practice, novice teachers can familiarize themselves with teaching routines and master basic teaching skills (lesson plan, implementation, and evaluation) smoothly and quickly.

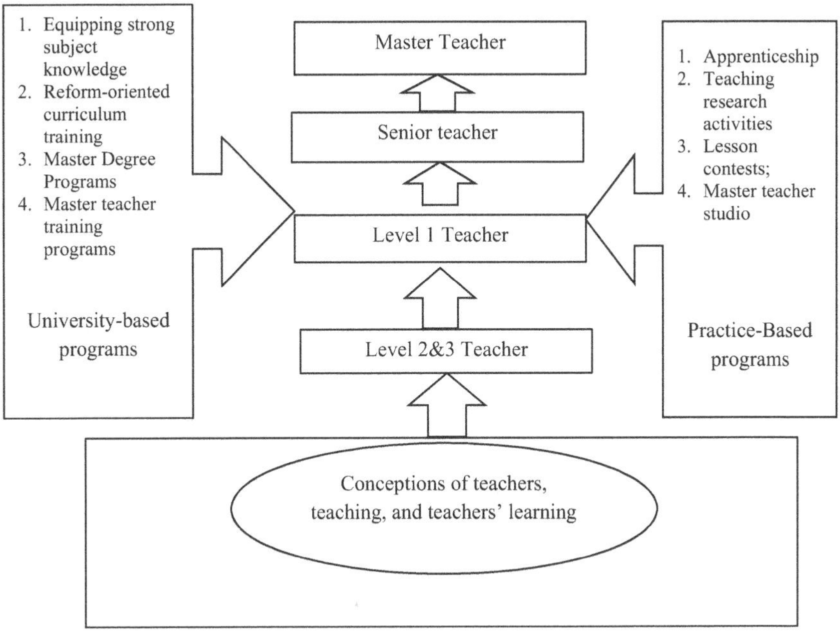

Fig. 2.1 A hybrid system of teacher professional development in China

When novice teachers become confident in teaching (level 2 and 3), various teaching research activities help them further develop their knowledge and skills for teaching and ability in conducting teaching research activities. As a result, some teachers get promoted as qualified/competent teachers (level 1). After that, teachers can further develop and demonstrate their professional competence through winning various contests (e.g., teaching lessons, explaining lessons or explaining problem solving) and doing teaching research projects. Thus, some of them could move up to be experts (senior level). Senior teachers have to take responsibility for mentoring young teachers. However, they can further develop to be *master teachers* (*full-senior*) in order to maximize their roles in mentoring other teachers, implementing new curricula and conducting teaching research on a large scale. The master teacher workstation and training programs meet such needs of teacher professional development.

2.4.2 *Discussion*

This well-structured, institutionalized teacher professional development system is deeply rooted in the Chinese conception of teachers, teaching, and teacher development. As argued by Leung (2003), emphasizing the subject knowledge of teachers is rooted in the conceptions of teachers and normal universities in China.

The graduates from normal universities (teachers' model) should be models for their students with regard to moral and academic aspects. It is necessary for teachers to have sound subject knowledge. Teaching in China is a publicly scrutinized enterprise (Li and Huang 2008). Giving open lessons, observing lessons, evaluating lessons, and reflecting on lesson are common components of teaching research activities. All teaching research activities focus on addressing teachers' practical concerns, solving their daily problems, and pursing exemplary lessons. In Chinese culture, it is believed that teachers can develop their knowledge for teaching "from examples and by doing" (Li et al. 2011a, b). Respecting seniors and learning from others, as indicated by "there are tutors among three people" (Confucian, The Analects), is a traditional trait. It is also believed that there are some good teaching models, which can be adapted (Cheng 2004). Thus, watching others' teaching and modeling good lessons are legitimate ways to develop teacher professional knowledge. Moreover, the ranking and promoting system provides mechanisms and incentive for teachers' continuous growth, while teaching research activities provide supportive platform for teachers to pursue their excellence.

Like teaching, teacher learning is a cultural activity (Stigler and Hiebert 1999). When other education systems reflect on what could be learned from the practices in China, it is important to identify their cultural values about teaching and teacher learning.

References

Bednarz, N., Fiorentini, D., & Huang, R. (Eds.). (2011). *International approaches to professional development for mathematics teachers*. Canada: Ottawa University Press.

Cheng, K. M. (2004). China: Turning the bad master into a good servant. In I. C. Rotberg (Ed.), *Balancing change and tradition in global education reform* (pp. 3–18). Lanham, Maryland: Scarecrow Education.

Fan, L., Wong, N. Y., Cai, J., & Li, S. (Eds.). (2004). *How Chinese learn mathematics: Perspectives from insiders*. Singapore: World Scientific.

Gu, L. (2005). *Reconstructing the culture of teacher's learning*. Retrieved December 12, 2015 from http://vod.js.edu.sh.cn/data/2005/0307/article_394.php (in Chinese).

Gu, L., & Wang, J. (2003). Teacher professional development through action education. *Curriculum, Teaching Materials and Pedagogies, 231*(1), 9–26. (in Chinese).

Huang, R., & Bao, J. (2006). Towards a model for teacher professional development in China: Introducing *keli*. *Journal of Mathematics Teacher Education, 9*, 279–298.

Huang, R., & Li, Y. (2008). Opportunities and challenges of the effectiveness of in-service mathematics professional development programmes in China. *Journal of Mathematics Education, 17*(3), 32–38. (in Chinese).

Huang, R., Li, Y., Zhang, J., & Li, X. (2011). Developing teachers' expertise in teaching through exemplary lesson development and collaboration. *ZDM—The International Journal on Mathematics Education, 43*(6–7), 805–817.

Huang, R., Peng, S., Wang, L., & Li, Y. (2010). Secondary mathematics teacher professional development in China. In F. K. S. Leung & Y. Li (Eds.), *Reforms and issues in school mathematics in East Asia* (pp. 129–152). Rotterdam: Sense.

Huang, R., Su, H., & Xu, S. (2014). Developing teachers' and teaching researchers' professional capacities in mathematics through Chinese lesson study. *ZDM—The International Journal on Mathematics Education, 46*(2), 239–251.

Huang, X., & Huang, R. (in press). Experienced teacher learning through master teacher workstation program: A case study. In Y. Li, & R. Huang (Eds.), *How Chinese teachers acquire and improve mathematics knowledge for teaching*. Rotterdam: Sense.

Lerman, S., & Zehetmeier, S. (2008). Face-to-face communities and networks of practicing mathematics teacher. In K. Krainer & T. Wood (Eds.), *Participants in mathematics teacher education: Individuals, teams, communities and networks* (pp. 133–155). Rotterdam: Sense.

Leung, F. K. S. (2003). Issues concerning teacher education in the East Asian region. *Asia-Pacific Journal of Teacher Education Development, 6*(2), 5–21.

Leung, F. K. S., & Li, Y. (Eds.). (2010). *Reforms and issues in school mathematics in East Asia*. Rotterdam, The Netherlands: Sense.

Li, S., Huang, R., & Shin, Y. (2008). Mathematical discipline knowledge requirements for prospective secondary teachers from East Asian perspective. In P. Sullivan & T. Wood (Eds.), *Knowledge and beliefs in mathematics teaching and teaching development* (pp. 63–86). Rotterdam, The Netherlands: Sense.

Li, Y., & Huang, R. (2008). *Developing mathematics teachers' expertise with apprenticeship practices and professional promotion system as contexts*. Paper presented at US—Sino Workshop on Mathematics and Science Education: Common Priorities that Promote Collaborative Research. June 22–June 27, 2008, Murfreesboro, TN.

Li, Y., & Huang, R. (2012). *How Chinese teach mathematics and improve teaching*. New York: Routledge.

Li, Y., & Li, J. (2009). Mathematics classroom instruction excellence through the platform of teaching contests. *ZDM-International Journal on Mathematics Education, 41*, 263–277.

Li, Y., Huang, R., Bao, J., & Fan, Y. (2011a). Facilitating mathematics teachers' professional development through ranking and promotion practices in the Chinese mainland. In N. Bednarz, D. Fiorentini, & R. Huang (Eds.), *International approaches to professional development of mathematics teachers* (pp. 72–87). Canada: Ottawa University Press.

Li, Y., Tang, C., & Gong, Z. (2011b). Improving teacher expertise through master teacher work stations: a case study. *ZDM—The International Journal on Mathematics Education, 43*, 763–776.

Liang, S. (2011). Open-class: An important component of teacher's in-service training in China. *Education, 1*(1), 1–5.

Liang, S., Claz, S., Defranco, T., Vinsonhaler, C., Grenier, R., & Cardetti, F. (2013). An examination of the preparation and practice of grades 7–12 mathematics from the Shangdong province in China. *Journal of Mathematics Teacher Education, 16*, 149–160.

Ma, L. (1999). *Knowing and teaching elementary mathematics: Teachers' understanding of fundamental mathematics in China and the United States*. Mahwah, NJ: Erlbaum.

Ministry of Education of China. (1994). The teachers ACT in P. R. China. Retrieved December 12, 2015. http://www.moe.edu.cn/s78/A02/zfs__left/s5911/moe_619/tnull_1314.html (in Chinese).

Ministry of Education of China. (1995). *Regulation of teachers' qualification*. Retrieved December 12, 2015 from http://www.moe.edu.cn/s78/A02/zfs__left/s5911/moe_620/tnull_3178.html (in Chinese).

Ministry of Education of China. (1999a). *Education revival agenda for 21st century*. Retrieved December 12, 2015 from http://www.moe.edu.cn/jyb_sjzl/moe_177/tnull_2487.html (in Chinese).

Ministry of Education of China. (1999b). *Agenda on deepening educational reform and entirely advancing quality education by state department of P. R. China*. Retrieved December 12, 2015 from http://www.moe.edu.cn/jyb_sjzl/moe_177/tnull_2478.html (in Chinese).

Ministry of Education of China. (2000). *Guidance for national training of key teachers at primary and secondary schools*. Retrieved December 12, 2015 from http://www.gxjs.com.cn/xmjz/guopei/zdyj.htm (in Chinese).

Ministry of Education of China. (2001a). *Mathematics curriculum standard for compulsory education stage (experimental version)*. Beijing: Beijing Normal University Press. (in Chinese).
Ministry of Education of China. (2001b). *The second national training program for backbone teachers at primary and secondary school, Intermediate report*. Retrieved December 12, 2015 from http://www.elab.org.cn.explore/evaluations/second/primary%20report.pdf (in Chinese).
Ministry of Education of China. (2011). *Syllabi of national teacher qualification examination*. Retrieved December 12, 2015 from http://www.ntce.cn/a/kaoshitongzhi/bishibiaozhun/ (in Chinese).
Ministry of Education of China. (2013a). *Interim rules for teacher qualification examination*. Retrieved December 12, 2015 from http://www.moe.gov.cn/publicfiles/business/htmlfiles/moe/s7711/201309/xxgk_156643.html (in Chinese).
Ministry of Education of China. (2013b). *Interim rules for regular registration of teacher qualification certificate*. Retrieved December 12, 2015 from http://www.moe.gov.cn/publicfiles/business/htmlfiles/moe/s7711/201309/xxgk_156643.html (in Chinese).
Ministry of Education, P. R. China. (2004). *Guidance guideline on further enhancing the teacher training for new curriculum at basic education*. Retrieved December 12, 2015 from http://www.edu.cn/zong_he_816/20060323/t20060323_110710.shtml (in Chinese).
Ministry of Human Recourses and Social Security and Ministry of Education of China. (2015a). *Guidance for deepening the reform of secondary and primary school teachers' promotion system*. Retrieved December 12, 2015 from http://www.mohrss.gov.cn/SYrlzyhshbzb/ldbk/rencaiduiwujianshe/zhuanyejishurenyuan/201509/t20150902_219575.htm (in Chinese).
Ministry of Human Recourses and Social Security and Ministry of Education of China (2015b). *Timeline for implementing the reform of secondary and primary school teachers' promotion system*. Retrieved December 12, 2015 from http://www.mohrss.gov.cn/SYrlzyhshbzb/dongtaixinwen/buneiyaowen/201509/t20150908_219977.htm (in Chinese).
Mullis, I. V. S., Martin, M. O., Foy, P., & Arora, A. (2012). *TIMSS 2011 international results in mathematics*. Chestnut Hill, MA: TIMSS & PIRLS International Study Center, Lynch School of Education Boston College.
National Mathematics Advisory Panel. (2008). *Foundations for success: The final report of the National Mathematics Advisory Panel*. Washington, D.C.: U.S. Department of Education.
OECD (2010). *PISA 2009 results: What students know and can do—student performance in reading, mathematics and science* (Vol. I). Retrieved December 12, 2015 from doi:10.1787/9789264091450-en
Peng, A. (2007). Knowledge growth of mathematics teachers during professional activity based on the task of lesson explaining. *Journal of Mathematics Education, 10*, 289–299.
Quan, L. (2009). Teacher professional growth within the context of master work station: A perspective of professional community. *Modern Education Science, 13*, 31–34. (in Chinese).
Schmidt, W., Tatto, M. T., Bankov, K., Blomeke, S., Cedillo, T., Cogan, L., et al. (2008). *The preparation gap: Teacher education for middle school mathematics in six countries*. East Lansing, MI: Michigan State University.
Stewart, V. (2006). China's modernization plan: What can US learn from China. *Education Week, 25*(28), 48–49.
Stigler, J. W., & Hiebert, J. (1999). *The teaching gap: Best ideas from the world's teachers for improving education in the classroom*. New York: Free Press.
Sullivan, P., & Wood, T. (2008). *The international handbook of mathematics teacher education* (Vol. 1): *Knowledge and beliefs in mathematics teaching and teaching development*. Rotterdam, The Netherlands: Sense.
Wang, J. (2009). *China mathematics education: Tradition and reality*. Jiangsu, China: Jiangsu Educational Press. (in Chinese).
Wang, J., & Paine, L. W. (2006). Mentoring as assisted performance: a pair of Chinese working together. *The Elementary School Journal, 102*, 157–181.
Wenger, E. (1998). *Communities of practice: Learning, meaning, and identity*. UK: Cambridge.
Wu, L., & Ge, J. (2015). Examination standard for school teacher qualification: Its background, purpose and content. *China Examination, 1*, 25–31. (in Chinese).

Yang, Y., & Ricks, T. E. (2012). Chinese lesson study: Developing classroom instruction through collaborations in school-based teaching research group activities. In Y. Li & R. Huang (Eds.), *How Chinese teach mathematics and improve teaching* (pp. 51–65). New York: Routledge.

Ye, L., Shi, H., & Zhang, J. (2011). Teachers' explanation of problem solving: An effective way of advancing teaching skills. *References of Secondary Mathematics Teaching, 4*, 55–57. (in Chinese).

Chapter 3
The Endless Long-Term Program of Mathematics Teacher Professional Development in Indonesia

Yaya S. Kusumah and Farida Nurhasanah

Abstract Like many developing countries, there are often competing demands for limited resources. In Indonesia, since 2005 there has been a concerted effort to improve the quality of education. As noted by the McKinsey's report, no education system can exceed the quality of its teachers one focus of change has been on teacher development, both at the pre-service and in-service levels. This chapter traces the chronological developments that have led to the national certification of teaching and the role of some agencies that have been involved in professional development programs for mathematics teachers. This chapter also describes some aspects of professional development for mathematics teachers that have emerged since the implementation of some government policies. The chapter concludes with some thoughts about professional development programs of mathematics teachers in the future.

Keywords Professional development · Mathematics teachers · PLPG · Pre-service teacher · In-service teachers · Lesson study · Realistic mathematics education · PMRI

3.1 Introduction

Indonesia is a developing country which has a large population of around 240 million people living on a land area of approximately 1.9 million square meters spanning from Sumatra to Papua. According to a report of the World Bank (2004),

Y.S. Kusumah (✉)
Department of Mathematics Education, Universitas Pendidikan Indonesia,
Bandung, Indonesia
e-mail: yayaskusumah@yahoo.com

F. Nurhasanah
Mathematics Education Department of Teachers Training and Education Faculty,
Sebelas Maret University, Surakarta, Indonesia
e-mail: Farida.nurhasanah@student.upi.edu

B. Kaur et al. (eds.), *Professional Development of Mathematics Teachers*, Mathematics Education – An Asian Perspective,
DOI 10.1007/978-981-10-2598-3_3

Indonesia has the largest population of school-age children in the world. Being a developing country coupled with a significantly large school-age population has resulted in issues related to education system. For a developing country to progress, a good education system is vital for developing human resource. In addition, according to the McKinsey's report (Barber and Mourshed 2007), no education system can exceed the quality of its teachers. Realizing that teachers are keys to student learning and an important part of the education system, the government of Indonesia has started to pay attention to teachers' quality. According to the Ministry of National Education and Culture (MoNEC 2008), 60 % of teachers in Indonesia did not have tertiary education comprising a 4-year bachelor degree or a 4-year diploma. Teachers were also paid relatively low salaries compared to other occupations (World Bank 2008). As teachers were not paid well compared to their counterparts in other professions, most good graduates from universities did not aspire to be teachers.

In 2005, the government launched a Law Number 14 about teachers and lecturers which is known as *Undang-Undang Guru dan Dosen* (UUGD) to alleviate problems in the teaching profession. This law mandates the standard qualifications of teachers suitable for teaching in schools. It also stipulates working hours of teachers, their duties as professionals, and their salaries. The UUGD appears to be a reform action. The UUGD (Depdiknas 2005) Number 14 Chapter IV unit 8 states that:

> Teachers must have academic qualifications, competencies (pedagogical, social and professional), national certification for teaching, good physical and spiritual health, and the desired ability to achieve national education goal (p. 6).

In the context of the above statement of the UUGD, the mandated academic qualification for teachers is at least a bachelor degree (S1) or a 4-year diploma (D4). This regulation ensures the recruitment of high-quality teachers and supports their professional development.

For the implementation of the abovementioned law, an elaborate and comprehensive plan for teacher professional development program, both for pre-service and in-service teachers, has been constructed. Figure 3.1 shows the plan formulated by MoNEC in 2012.

3.2 Pre-service Education of Mathematics Teachers

In Indonesia, before the 1990s, there were several institutions involved in pre-service education of teachers, namely School for Teacher of Type B (SGB), Upper School for Teachers (SGA), School for Teacher Education (SPG), Education Institution of Religion Teachers (PGA), Teacher School for Sports (STO), Graduate School of Education for Teacher Training and Educational Sciences (STKIP), and Institute for Teacher Training and Educational Sciences (IKIP). All these institutions were responsible for preparing prospective teachers for primary and secondary

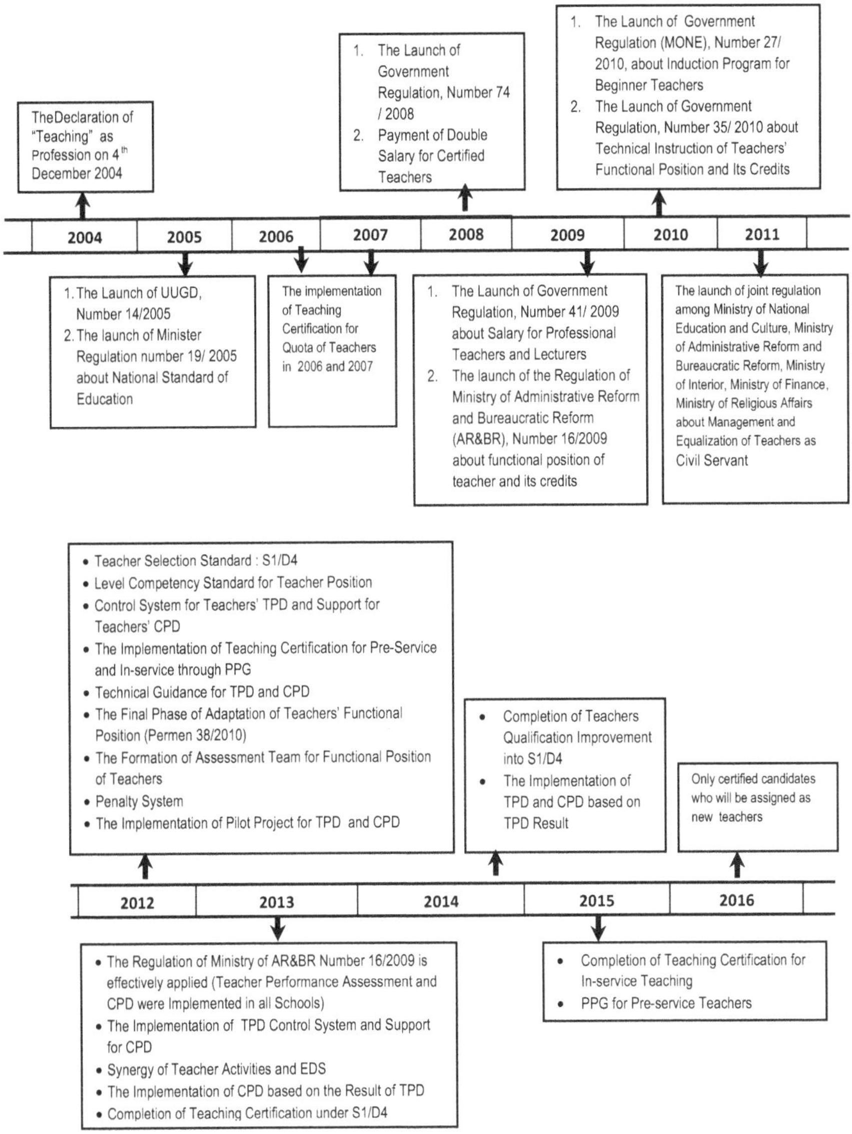

Fig. 3.1 Milestone of teachers professional development in Indonesia (MoNEC 2012)

schools. However, the situation was rather different before the 1970s when prospective student teachers were selected through a national scholarship program. Following a population explosion in Indonesia in the 1960s, that resulted in a sudden significant increase in school-going children, the government responded to the situation by initiating an expansion of the number of primary schools. This program was called SD Inpres (Jalal et al. 2009). As a consequence of this program, there was a huge demand for teachers, and this led to a recruitment process of prospective teachers which was less selective. Hence teacher recruitment was no longer exclusive for the best students. This change also encouraged many private institutions to be involved in pre-service teacher education.

Institutions that are responsible for preparing pre-service teachers in Indonesia are known as The Teachers Training Institution (LPTK). Since the 1990s there have been several institutions for teacher preparation such as: Faculty of Teacher Education and Educational Sciences (FKIP) in a university, Faculty of Mathematics and Science Education (FPMIPA) in Institute of Teacher Training and Educational Sciences (IKIP), Graduate School of Education for Teacher Training and Educational Sciences (STKIP), and Institute of Teacher Training and Educational Sciences (IKIP). Starting from 1998, 10 IKIPs have been expanded to include departments focused on basic sciences courses in addition to their existing departments of education and transformed into universities. The IKIP Bandung, has been transformed into Universitas Pendidikan Indonesia (UPI), IKIP Yogyakarta into Universitas Negeri Yogyakarta (UNY), IKIP Jakarta into Universitas Negeri Jakarta (UNJ), and likewise another 7 IKIPs have also become universities.

Before the UUGD was implemented, pre-service teacher programs varied in duration depending on which level teachers were preparing to teach. Now, all prospective teachers are required to have at least a bachelor degree (S1) or 4-year diploma (D4). They may obtain their degree or diploma from FKIP/FPMIPA in a university or STKIP. In addition, based on the new regulation concerning the national certification program, they have to undertake the Professional Education Program (PPG). This scheme is for all teachers, including mathematics teachers. Under this scheme, students from mathematics departments (in the faculty of mathematics and science, outside of FKIP/IKIP/STKIP) may also join the PPG and upon completion become mathematics teachers.

In 2013, there were 415 LPTKs in Indonesia, 38 of which were government institutions and the rest were private universities or STKIP (Hidayatullah 2013). Due to the large number of institutions, both government and private, there have been inconsistent standards resulting in issues related to the quality of teachers produced by the institutions. In order to address this challenge concerning mathematics teachers, the Indonesian Mathematical Society (IndoMS) initiated and formulated the standards of curriculum for mathematics education and mathematics departments for pre-service mathematics teachers. The IndoMS Team (2013) recommended that curriculum for mathematics and mathematics education must consist of at least 7 strands of study, they are: (1) general field of study (for example, science, humanities or knowledge subjects), (2) mathematics content field of study (for example, real analysis, abstract algebra, complex numbers, etc.), (3) school mathematics field of study (for example, school mathematics topics such as number, geometry, algebra, etc.), (4) mathematics education field of study (for example, learning theories of mathematics, use of teaching aids to develop conceptual knowledge, etc.), (5) pedagogy field of study (for example, psychology of teaching and learning, managing students, etc.), (6) additional skill field of study (for example, enrichment courses like academic writing, public communication, etc.), and (7) special field of study (this depends on the expertise of the university and their prime focus for example). Pre-service teachers need to take at least 144 credits in 8 semesters including teaching practice and thesis writing in the final semester.

3.3 National Teaching Certification for In-service Teachers that Lead to Teacher Professional Development

In Indonesia, one of the reform programs from the government related to the development of teachers is the National Teaching Certification program. This program is an outcome of the implementation of UUGD no 14 (Chapter IV Unit 18) which stipulates that all teachers in Indonesia must hold a national teaching certificate as a teaching license. This is part of the requirements they have to meet in order to be professional teachers.

The goal of the government is to certify all practicing teachers by 2015. As teachers in the past had varied academic qualifications, the implementation of this policy has been a challenge for the Government of Indonesia (GOI). Some schemes have been put in place to support this effort. In 2009, with the setting up of the Ministry of Administrative Reform and Bureaucratic Reform (KemenPAN-RB) that oversees the management of civil servants, a structure was also developed for the career advancement of teachers. Based on the Regulation, the positions and levels for teachers in Indonesia (PerMenPan Number 16 2009) were classified as follows:

- Beginning Teacher: level III/a and level III/b
- Intermediate Teacher: level III/c and level III/d
- Advanced Teacher: level IV/a, level IV/b and level IV/c
- Master Teacher: level IV/d and level IV/e.

The three different ways for in-service teachers of all subjects, including mathematics to get their certification, are as follows:

(1) **PSPL (Direct Scheme for Certification)**
This scheme is available only for two groups of teachers. The first group consists of teachers with academic qualifications of master degree (S2) or Ph. D. (S3) and teachers who have reached level IV/b in their teaching career. The second group comprises teachers who have reached level IV/c. These teachers qualify for the PSPL scheme. They only need to submit their documents for verification. If all these requirements have been fulfilled, they are entitled to have professional certificate. Those who are unsuccessful should take the Initial Competency Test. If they pass the test they can join the PLPG scheme (which is described later). Otherwise, they have to retake the test until they are successful.

(2) **Portfolio (PF)**
This scheme is available for three groups of teachers. The first group consists of teachers who have academic qualification S1 (university graduate) or D4 (4-year diploma). The second group covers all teachers who do not have S1/D4 qualification, but their age is at least 50-year old and they also have more than 20 years of teaching experience. The third group consists of teachers who have obtained rank of IV/a. These teachers can also choose the PLPG scheme. However, the teachers in the second group and the third group were only eligible for this scheme before the beginning of 2014. In this portfolio scheme, teachers need to write their own portfolio. In this scheme, portfolio refers to a collection of physical evidences that describe work experience/accomplishments achieved during professional duties as a teacher in a given interval of time. The documents of portfolio comprises 10 components, namely: (1) academic qualification; (2) training and workshops; (3) teaching experiences; (4) classroom management; (5) supervisor evaluation (6) academic achievement; (7) work of professional development; (8) participation in seminar forum; (9) organization experience in the field of social and education; (10) awards relevant to education field. The portfolio is submitted to the quality assurance agency of local government (LPMP) and assessed by selected education institutions (LPTK) which have been authorized to assess the documents in the portfolio. If the assessment result meets the Passing Grade (PG), the teachers will pass the certification process. If they do not pass the PG, then they have to join the Initial Competency Test. If they pass this test they can join PLPG, otherwise they have to upgrade their competency and wait for the next session.

(3) **PLPG (Professional Education and Training for Teachers)**
Another alternative for those two groups of teachers is PLPG scheme. Teachers who choose this scheme should take Initial Competency Test as a requirement. If they pass the test, they can join PLPG session for a period of time (100 h of face-to-face meeting). Finally at the end of the session they will have a final test. Only those who pass the test will get a professional teaching certificate. For those who fail have to improve their competency and retake the certification scheme.

3.4 National Teaching Certification for Pre-service Teachers

As most of the in-service teachers completed their certification process by the end of 2014, the PLPG program was replaced by the Professional Education for Teachers (better known as *Pendidikan Profesi Guru*—PPG). PPG is a professional development program for pre-service teachers to prepare graduates who have hold bachelor degree from education and non-education university and have talents and interests as teachers to master full teacher competencies based on national standards to be able to obtain certificate of professional educators in early childhood education, primary education, and secondary education (Regulation of MoNEC No 87, 2013). This program is compulsory for graduates, from universities which conduct pre-service teacher education and also other university courses that are relevant for teachers such as Pure Mathematics. The duration of the program is 1 year or 1.5 years (consisting of two or three semesters) depending on the performance of the pre-service teachers. In the first semester, pre-service teachers learn pedagogical concepts and in the second semester they engage in teaching practice. The program is conducted only at a number of LPTKs designated by Ministry of National Education and Culture (MoNEC 2014).

There are two types of PPG program for pre-service teachers: (1) General PPG and (2) PPG-SM3T. General PPG can be enrolled either by students who were graduated from LPTK or non-LPTK. All participants from this scheme have to pay tuition fee and have to join one-year matriculation program beforehand. On the other hand, PPG-SM3T is prepared only for pre-service teachers who have spent a one-year SM3T program (*Program Sarjana Mendidik di Daerah Terluar, Terdepan, dan Tertinggal*). SM3T program is initiated by MoNEC for sending teachers to isolated, outermost, and underdeveloped regions in Indonesia for one year. This program is devoted special only for students who were graduated from LPTK. PPG participants from this scheme do not need to pay tuition fee and they have to stay in dormitory during the program. Pre-service mathematics teachers who want to take this scheme are the best candidates for mathematics teachers because they have to follow selection test. This is one of many efforts conducted by government to obtain the best students as candidate for teacher as well as to overcome problems related to accessibility of teacher in isolated, outermost, and underdeveloped region in Indonesia.

Teachers having completed a four-year bachelor or diploma courses followed by the PPG certification process are considered as competent teachers, so they are entitled to hold teaching certificate and admitted as professional teachers. The teachers, however, should realize that pre-service education and certification is not an end in itself. To keep abreast of advances in knowledge and technology they must continuously develop themselves. Hence Teacher Professional Development (TPD) is an important aspect of their career development paths. The government needs to guarantee that the professionalism of teachers must be sustainable. Therefore, the endless long-term program of professional development of teachers must be the next step.

3.5 Professional Development of Mathematics Teachers

Teacher professional development, particularly for mathematics teachers, is a complex process. Its development and enhancement is an endless long-term program which needs a variety of efforts, carried out by experts, principals, supervisors, curriculum developers, and the teachers themselves. It may also take various forms. Meaningful professional development often results from autonomous activities chosen by the teachers themselves as they seek ways of understanding and teaching mathematics. Research has shown that externally imposed professional development activities, although well intentioned, are often doomed to failure (Castle and Aichele 1994).

Although teachers are inducted into their profession during pre-service education, it may be said that their professional development mostly takes place during their in-service years. There are often two main types of programs, which teachers participate in for their professional development. The first type is one where teachers engage in learning while being away from their teaching duties at school and the second type is where they engage in learning while carrying on with their teaching duties at school. Programs of the first type are courses of further study at institutions of higher learning, attending in or out-of-school training, seminars, workshops, and conferences which are organized for fostering new understanding towards mathematics teaching and for formulating new perspective in teaching mathematics (Jones et al. 1994).

There are several agencies and also approaches that facilitate the professional development of mathematics teachers, in Indonesia. The two main institutes that provide professional development programs for mathematics teachers are the Educational Institute of Quality Assurance (*Lembaga Penjaminan Mutu Pendidikan*-LPMP) and Institute for Mathematics Teacher Training (P4TK-Matematika). LPMP is a local province institution available for mathematics teachers in a region, whereas P4TK-Matematika is a national institution that is located in Yogyakarta.

Due to the large number of mathematics teachers, 1.17 million in 2012, the two main institutes are unable to serve the needs of all mathematics teachers in the country (Wijaya 2013). Therefore other than the LPMP and P4TK-Mathematika there are also some institutions which engage in the professional development programs for mathematics teachers. In some regions such as West Java, Jakarta, Central Java, Makasar, there are working groups for Mathematics Teachers (better known as *Musyawarah Guru Mata Pelajaran Matematika*—MGMP Matematika), which have monthly meetings and frequent seminars on mathematics and mathematics education for mathematics teachers. MGMP is a professional working group for mathematics teachers in local districts, which is usually led by key teachers who also organize activities and periodic meeting in the group. Study by Evans et al. (2009) shows that MGMPs still need to struggle in enhancing the quality of mathematics teachers, in subject matters and pedagogy.

Among those institutions and stakeholders which involve in enhancing teachers quality in Indonesia, the most comprehensive engagement in teacher professional development both for pre-service and in-service is dominated by teacher training institutions (LPTKs) (Evans et al. 2009). LPTK prepares pre-service teachers to be professional teachers as well as providing various programs of professional development for in-service mathematics teachers. PLPG and PPG programs are also organized by LPTKs (together with their qualified counterparts) that meet qualification on behalf of the Minister of MoNEC.

3.6 Professional Development Programs for In-service Mathematics Teachers in Indonesia

Long before the implementation of UUGD, TPD programs for mathematics teachers have been organized and conducted by either government or nongovernment institutions. TPD for mathematics teachers in Indonesia can be categorized into two types, namely top-down type and bottom-up type. Top-down TPD programs are initiated by the government and conducted by LPTK, P4TK, and LPMP, whereas the bottom-up TPD programs are mostly initiated and run by non-government agencies or groups of researchers.

According to Evans et al. (2009), there are two pathways of TPD for in-service teachers in Indonesia: the traditional pathway and the newer pathway. The traditional pathway consists of three types: (1) full university attendance to complete additional academic study; (2) Distance learning through the Open University (*Universitas Terbuka*); and (3) attendance in courses conducted by national agencies, such as LPMP or P4TK-Matematika. The newer type which is available for mathematics teachers are: (1) distance learning program held by LPTK; (2) local in-service activities provided by MGMP or KKG (Working Group for Teachers); (3) scholarship programs from the government; (4) noncredit workshops; and (5) workshops for socialization of government policies.

Both types of pathways have been available even before the implementation of UUGD. In 2006, more than 60 % of the total 2.78 million in-service teachers have not reached the level of academic qualification of a four-year bachelor degree (S1/D4) (MoNEC 2008). Before the implementation of UUGD, most of the teachers had insufficient financial support for upgrading their qualification or attending professional development program. In addition, due to the huge number of teachers and geographic aspect, most of the teacher professional development programs could only be reached by small number of teachers.

To overcome these problems, the government of Indonesia (GOI) launched scholarship program for mathematics teachers who want to pursue master degree in a university inside the country. Teachers who have passed the selection will be encouraged to finish their study without the obligation of teaching during their study. This program, designed specially for teachers who are also civil servants under MoNEC, is known as Scholarship Program of P2TK (*Pembinaan Pendidik dan Tenaga Kependidikan*). The scheme is available for teachers from primary until high schools.

Another scholarship that is also available for mathematics teachers is the one which was launched by the Government of Indonesia under the scheme designed and implemented by LPDP (*Lembaga Pengelola Dana Pendidikan*), a national agency that organizes educational fund. LPDP enables mathematics teachers to apply for scholarship scheme to pursue master degree either overseas or in the country.

Between 2003 and 2013, the government of Indonesia initiated schools to acquire the status of an International Standard School, by encouraging mathematics and science teachers to teach their subjects in English. This was intended to improve students' achievement in mathematics and science, so that they could be at par with other students in developed countries. As part of this initiative many LPTKs were involved in training teachers to teach Mathematics and Science in English (Setyorini and Sofwan 2011).

Yet another initiative, focused on models of teaching known as the "PAKEM" (stands for active, creative, effective, and joyful instruction) initiated by the Educational Institute of Quality Assurance (LPMP) has also provided primary school mathematics teachers with professional development. Meanwhile, Teacher Training Institutions (LPTKs) have been involving in providing in-service teacher development program. These programs are mainly conducted by lecturers as researchers whilst collaborating with other education institutions.

Apart from the above mentioned information, the Japan International Cooperation Agency (JICA) has also been very active in the training of mathematics (and other subject matter) teachers for strengthening the quality of their teaching. The first program launched by JICA was a pilot program involving 2 schools in Bandung (West Java), 2 schools in Yogyakarta (Yogyakarta City and Central Java), and 2 schools in Malang (East Java). In this project, 3 universities were involved, namely: Universitas Pendidikan Indonesia (UPI), Universitas Negeri Yogyakarta (UNY), and Universitas Negeri Malang (UM). This program has evolved into the Lesson Study (Fernandez and Yoshida 2004) type of professional development for enhancing mathematics teachers' competency and professionalism. The outcome of this program has resulted in many mathematics teachers changing their style of teaching, from teacher-centered to student-centered. They also began to apply multi-way traffic communication and act as facilitators and motivators rather than instructors or information deliverers (Kusumah and Asep 2008; Kusumah 2008a, b, 2016). At present, Lesson Study is a mode of professional development in many provinces throughout Indonesia. In addition, in Yogyakarta, Lesson Study is one of the training courses conducted by SEAMEO-QITEP in Mathematics, for upgrading mathematics teachers' competency in ASEAN countries.

In addition to the professional development programs for in-service teachers described in the previous paragraphs there is one TPD which may be unique to in-service mathematics teachers in Indonesia. This TPD, which is of the noncredit type, is part of the movement to adapt Realistic Mathematics Education (RME) in Indonesia to enhance the quality of teaching and learning of mathematics (Hadi 2002; Sembiring et al. 2010; Ekholm and van den Hoven 2009). This movement is called "PMRI" (*Pendidikan Matematika Realistik Indonesia*).

Teachers (in this movement) are trained to teach mathematics based on realistic mathematics education contexts through various workshops. The institute also collaborates with other bodies such as the P4TK-Matematika and universities in providing professional development for mathematics teachers. PMRI promotes a bottom-up movement to enhance the quality of mathematics education in Indonesia. It was started in 2001 involving 4 universities (UPI, USD, UNY, and UNESA) and 12 primary schools in Java Island. Now this movement has spread to other islands in Indonesia, such as Sumatra and Kalimantan. Widjaja and Dolk (2010) studied the implementation of RME and found that it can be used as a way to build, support, and enhance teachers' capacity to foster mathematics learning. They found that teachers also changed their learning process, from telling to facilitating learning. The effectiveness of the movement has spurred it on from primary schools to junior high schools.

Due to the huge number of mathematics teachers throughout Indonesia, in urban areas as well as in rural and remote areas, unfortunately the program of the two projects (RME/PMRI and Lesson Study) could not be reached by all mathematics teachers. However, by the support from certification program after the implementation of UUGD, mathematics teachers who have not been touched by Lesson Study and Realistic Mathematics Education trainings and seminars, actively seek the latest information and development which they access through their own strategies: individual learning, distance learning, or on line learning. They utilize all media of information and communication technology for enhancing and upgrading their knowledge, skills, and horizon.

3.7 Conclusion

The launch of UUGD followed by a number of government regulations has definitely enthused teachers to engage in Continuous Professional Development (CPD) so as to develop their teaching career. Under these regulations, teachers have to join CPD during their teaching career, which foster them to publish a paper or to create an innovation as the requirement for their promotion to the higher career level. Definitely, this condition encourages mathematics teachers, who work in the front line of education, to do independent CPD through various pathways.

The implementation of UUGD and the scheme of certification program have certainly influenced the teachers' interest, motivation, and enthusiasm in teaching and learning process, which in turn maximize their performance in doing their tasks. The indirect impacts of the UUGD implementation, is that certified mathematics teachers nowadays have wide opportunity to gradually enhance their capacity and capability in teaching, as they become IT/ICT literate, and have easiness in searching important and current relevant information from the Internet with sufficient devices at their hands. This easiness is due to the sufficient financial fund they obtain under the scheme of certification program.

In addition, the availability of sufficient communication devices, Internet and network, together with teachers' literacy in information and communication technology, are the main factors which foster teacher professional development. In the long run, they will be able to strengthen the competitiveness of Indonesian human resources and put Indonesia in equal position with other countries.

It is also apparent that the national certification of teachers is only the first step in ensuring that the baseline standards of teachers are aligned. It is imperative for in-service teachers, as the agent of change, to engage in lifelong learning and therefore teachers must have intrinsic motivation to continuously enhance their knowledge and skills. In addition, their career advancements must be also tagged onto their continuous development.

References

Barber, M., & Mourshed, M. (2007). *McKinsey report: How the world's best-performing school systems come out on top*. McKinsey & Company.

Castle, K., & Aichele, D. B. (1994). Professional development and teacher autonomy. In D. B. Aichele & A. F. Coxford (Eds.), *Professional development for teachers of mathematics* (pp. 1–8). Reston, Virginia: National Council of Teachers of Mathematics Inc.

Depdiknas (Department of National Education). (2005). *Undang-Undang Republik Indonesia* No.14 tahun 2005 (Law of Republic Indonesia about Teachers and Lecturers). Jakarta: Departemen Pendidikan Nasional.

Ekholm, M., & van den Hoven, G. H. (2009). *PMRI - Majulah!* Naarden: Zet&Print.

Evans, D., Tate, S., Navarro, R., & Nicolls, M. (2009). *Teacher education and professional development in Indonesia: A gap analysis*. USAID, Aguirre Division of JBS International, Inc. Retrieved June 3, 2016.

Fernandez, C., & Yoshida, M. (2004). *Lesson study: A Japanese approach to improving mathematics teaching and learning*. London: Lawrence Erlbaum Associates, Publishers.

Hadi, S. (2002). *Effective teacher professional development for the implementation of realistic mathematics education in Indonesia*. Enchede: University of Twente.

Hidayatullah, F. (2013). *Prof. Furqon: Dari 415 LPTK hanya 38 yang Negeri*. Available online at: http://www.timlo.net/baca/68719507150/prof-furqon-dari-415-lptk-hanya-38-yangnegari/

Jalal, F., Samani, M., Chang, M. C., Stevenson, R., Bagatz, A. B., & Negara, S. D. (2009). *Teacher certification: A strategy for teaching quality improvement*. MoNEC: Jakarta.

Jones, A. G., Lubinski, C. A., Swafford, J. O., & Thornton, C. A. (1994). A framework for the professional development of K-12 mathematics teachers. In D. B. Aichele & A. F. Coxford (Eds.), *Professional development for teachers of mathematics* (pp. 23–36). Reston, Virginia: National Council of Teachers of Mathematics Inc.

Kusumah, Y. S. (2008a). *Enhancing students' mathematical learning through teacher professional development*. Paper presented at the Third International Conference on Mathematics and Statistics: Bridges for Academia, Business, and Government in the Entrepreneurial Era, held by Moslems Statisticians and Mathematicians Society East Asia (MSMSSEA), Department of Mathematics and Department of Statistics of IPB, and Department of University Malaysia Terengganu, August 5–6, 2008.

Kusumah, Y. S. (2008b). *Enhancing students' mathematical learning through teacher professional development*. Paper presented in the Third International Conference on Mathematics and Statistics, held in Bogor Agricultural University (IPB), Bogor-Indonesia, 5–6 August, 2008.

Kusumah, Y. S. (2016). *The implementation of lesson study in ASEAN countries (an endless long-term program for enhancing teacher professionality)*. Paper for In-service Mathematics Teacher Training Program in SEAMEO QITEP in Mathematics, Yogyakarta.

Kusumah, Y. S., & Asep S. H. (2008). *Improving mathematics teaching and learning: Best ideas and experiences from the implementation of lesson study (a case study in East Sumedang District)*. Paper presented in the First International Conference on Lesson Study. Bandung: FPMIPA—Indonesia University of Education.

MoNEC. (Ministry of National Education). (2008). *Monitoring and evaluation of the implementation of the certification of inservice teachers*. Jakarta: Konsorsium Sertifikasi Guru (KSG).

MoNEC. (2014). *Sertifikasi Guru dalam Jabatan: Buku 1 Pedoman Penetapan Peserta*. Jakarta: BPSDM-PMP.

MoNEC. (2012). *Kebijakan Pengembangan Profesi Guru*. Available online at: http://repository.ung.ac.id/get/kms/2887/Materi-KEBIJAKAN-PENGEMBANGAN-PROFESI-GURU-presentasi.pptx

PerMenPAN Number 16. (2009). *Peraturan Menteri Pendayagunaan Aparatur Negara Number 16*. Indonesia: Ministry for the Empowerment of State Apparatus.

Sembiring, R., Hoogland, K., & Dolk, M. (2010). *A decade of PMRI in Indonesia*. APS: Utrecht.

Setyorini, A., & Sofwan, A. (2011). Teaching mathematics and science in English in pilot international standard high school in Indonesia. *Language Circle Journal of Language and Literature, 5*(2), 81–86.

Team, Indo M S. (2013). *Laporan Rekomendasi Capaian Pembelajaran serta Struktur Kurikulum Minimal Prodi S1 Matematika, Pendidikan Matematika, Statistika dan Ilmu Komputer/Teknik Informatika*. IndoMS: Yogyakarta.

Widjaja, W., & Dolk, M. (2010). Building, supporting, and enhancing teachers' capacity to foster mathematical learning: Insights from Indonesian classroom. In Y. Shimizu, Y. Sekiguchi & K. Hino (Eds.), *the Proceedings of the 5th East Asian Regional Conference on Mathematics Education* (*EARCOME*) (Vol. 2, pp. 332–339). Tokyo: ICMI—International Commission on Mathematical Instruction. August 18–22, 2010.

Wijaya, A. (2013). Upaya Pemerataan Akses Pendidikan Melalui Berbagai Bentuk Kegiatan Diseminasi Hasil Diklat. Available onine at: http://p4tkmatematika.org/file/ARTIKEL/Artikel%20Pendidikan/DISEMINASI%20DIKLAT.pdf

World Bank. (2004). *Education in Indonesia: Managing the transition to decentralization* (vol. 1). Report Number 29506.

World Bank. (2008). *Teacher employment and deployment in Indonesia: Opportunities for equity, efficiency and quality improvement*. Washington, DC: World Bank.

Chapter 4
Lesson Study: The Fundamental Driver for Mathematics Teacher Development in Japan

Akihiko Takahashi

Abstract How do Japanese teachers develop knowledge and expertise for teaching mathematics effectively? Their journey begins while in university attending various teacher preparation programs. This undertaking does not end once they become teachers. They are expected to be lifelong learners to become effective educators. Lesson Study has been the fundamental driver of improvement in teaching and learning in Japan. This chapter describes how Lesson Study supports teachers in their continuous growth to become effective teachers of mathematics and provides empirical evidence based on current research projects conducted by the author as a part of Project IMPULS at Tokyo Gakugei University.

Keywords Lesson study · School-wide · Research steering committee · Knowledgeable other · Lifelong learning

4.1 Introduction

It is obvious that teachers cannot teach content beyond their knowledge (National Mathematics Advisory Panel 2008), but knowledge of content is not enough to teach effectively. Japanese mathematics educators and teachers identify three levels of expertise of mathematics teaching (Sugiyama 2008):

Level 1: The teacher can tell students the important basic ideas of mathematics such as facts, concepts, and procedures

Level 2: The teacher can explain the meanings and reasons of the important basic ideas of mathematics in order for students to understand them

A. Takahashi (✉)
DePaul University, College of Education, Chicago, USA
e-mail: atakahas@depaul.edu

A. Takahashi
Project IMPULS at Tokyo Gakugei University, Tokyo, Japan

B. Kaur et al. (eds.), *Professional Development of Mathematics Teachers*, Mathematics Education – An Asian Perspective,
DOI 10.1007/978-981-10-2598-3_4

Level 3: The teacher can provide students with opportunities to understand these basic ideas, and support their learning so that the students become independent learners

(Trans. Takahashi 2011)

Level 1 teaching does not require any special knowledge. In fact, having received decent grades in mathematics in grade school may be all that is necessary. But for Level 2 and Level 3 teaching, special knowledge and expertise are required. Sugiyama (2008) writes that during the early twentieth century, which is early in the evolution of the Japanese public education system, most elementary school teaching was at Level 1. Instructors simply told their students the facts and expected them to memorize those facts through practice, and contemporary textbooks were designed to support this form of instruction. Certainly it is important for teachers to be able to tell students basic facts, but today in Japan a teacher must provide instruction at Level 2 or 3 to be considered a professional.

To teach at Level 2, one must possess knowledge of mathematics beyond what is needed in everyday life or what is required to solve problems in school textbooks. For example, knowing the "invert and multiply" rule for division of fractions is enough to be a Level 1 teacher, but for Level 2 a teacher must be able to explain how multiplying by the reciprocal of a fraction produces the quotient. This type of knowledge is important for helping students understand mathematics (e.g., Ball et al. 2008). But while Level 2 is considered professional teaching, Japanese mathematics educators believe that all mathematics teaching should be at Level 3, because they have seen that Level 2 teaching does not enable students to develop mathematical proficiency with understanding.

A majority of current government-authorized mathematics textbooks in Japan are designed to support Level 3 teaching. These textbooks are designed for teachers to present students with a problem that the students have not yet learned how to solve. The texts provide structure and allow the teachers to guide the conversation in such a way that students can arrive at a new understanding as a result of their own efforts in solving the new problem. The philosophy behind Level 3 teaching is that students should be given a reasonable amount of independent work, such as problem-solving, in order to develop the knowledge, the understanding, and the skills of mathematics (National Research Council 1989; Polya 1945).

Japanese mathematics educators can safely assume that most university students have knowledge of mathematics for Level 1 teaching. Their concern, therefore, is to move those students toward Level 2. But there is not enough time in the preservice program to equip the future teachers with Level 2 knowledge of all the contents they might be required to teach. So, Japanese universities focus on training students to acquire Level 2 knowledge through a careful study of teaching materials (Sugiyama 2008). They offer courses for elementary mathematics teacher preparation that focus mainly on examining the contents of mathematics for elementary grades and developing a deeper understanding of these contents. This process is called *kyouzai kenkyuu*—in other words, “studying teaching materials for establishing deeper understanding for better teaching” (Watanabe et al. 2008).

For example, there are several formulas for finding the area of basic geometric figures. Most students who come to a teacher preparation program already know those formulas and can use them to find the area of basic figures. Using contents from published textbooks for elementary grades, university courses help the prospective teacher see how the formulas are developed, how they are related to each other, how they are related to other areas in mathematics, and potential difficulties students might have with learning the formulas. Investigating a topic in this way is typical of *kyouzai kenkyuu* and is an essential part of teachers’ preparation for everyday teaching; hence these courses also introduce the prospective teachers to *kyouzai kenkyuu* as a critical step in preparing lessons.

Preparing student teachers for Level 3 teaching is even further beyond the scope of what can be accomplished during the teacher preparation programs at the university. Japanese educators believe that teachers cannot master Level 3 teaching simply by listening to lectures, reading textbooks, and watching videos. Learning to teach at Level 3 is demanding and time-consuming, and a career-long process. But the universities do help prospective teachers understand what Level 3 teaching is, and teaches them the pathway to it.

4.2 Helping Practicing Teachers Increase Their Knowledge and Expertise

When designing professional development programs for practicing teachers, it is useful to recognize that professional development falls into two categories; phase 1 and phase 2. Phase 1 professional development (phase 1 PD) focuses on increasing a teacher’s knowledge for teaching mathematics, while phase 2 professional development (phase 2 PD) focuses on developing expertise for teaching mathematics—that is, the ability to use new knowledge in the classroom (Takahashi 2011).

Moving from Level 1 to Level 2 can be achieved through phase 1 PD, and most university courses in teacher preparation programs, which may include reading books, listening to lectures, and observing well-designed mathematics classes, fall into the category of phase 1. Practicing teachers may need phase 1 PD from time to time to update their knowledge for teaching. On the other hand, Level 3 teaching requires very different classroom practices and skills than Level 2 teaching, and learning these practices and skills requires phase 2 PD. To develop this expertise requires considerable teaching experience with a reflection component. Japanese teachers and researchers use Lesson Study to develop the deeper knowledge and the expertise necessary to make Level 3 teaching available for their students.

4.3 Lesson Study as a Fundamental Driver for Mathematics Teacher Development

Lesson Study has been the primary mechanism of professional development for both prospective teachers and practicing teachers since the Japanese public education system started (Lewis 2000; Lewis and Tsuchida 1998a; Makinae 2010; Murata and Takahashi 2002; Takahashi 2000; Takahashi and Yoshida 2004; Yoshida 1999a). In Lesson Study, teachers conduct intensive *kyouzai kenkyuu*—study the standards, read relevant research articles, examine available curricula, and other materials—and work together to design a lesson focused on a problematic topic while also addressing a broader research theme related to teaching and learning. The lesson they design, known as a "research lesson" (*kenkyuu jugyou*), is taught by one teacher from the planning team while the other team members observe. The planning team and observers then conduct a post-lesson discussion focusing on how students responded to the lesson in order to gain insights into the teaching–learning process. Figure 4.1 illustrates the typical process of Lesson Study.

4.3.1 Introduction to Lesson Study During Teacher Preparation

Japanese teachers acquire first-hand experience of Lesson Study during their student teaching. Each student teacher has the opportunity to carefully observe lessons taught by the cooperating teacher and by other student teachers. Based on the observation of the lessons, student teachers write a detailed lesson plan and teach the lesson based on that lesson plan. After each lesson is taught, the cooperating teacher, the student teacher who taught the lesson, and other student teachers who observed the lesson have a mini-version of a post-lesson discussion. This is based

Fig. 4.1 Lesson study cycle to impact student learning

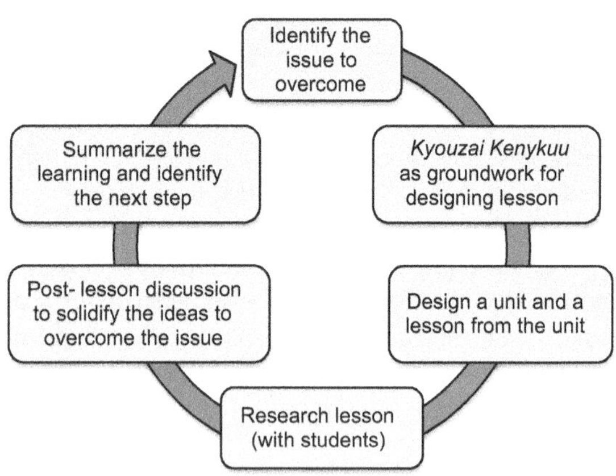

on the careful observation of the students during the lesson. At the end of the student teaching, the school conducts a formal research lesson for the student teachers by having other teachers at the school observe them teach; this research lesson is an initiation of the student teachers into the teaching profession. Through this experience of practicing Lesson Study during student teaching, each prospective teacher learns the basics of Lesson Study. For example, how to observe students during lessons, how to prepare a lesson plan for a research lesson, how to be part of the post-lesson discussion, and how to write a summary report of a Lesson Study cycle.

4.4 An Example of School-Wide Lesson Study to Support Teachers Becoming Life-Long Learners

During Lesson Study, teachers have the opportunity to look closely at teaching practices and to judge, based on student learning, whether the lesson properly supports the students in learning mathematics. Researchers credit Japanese Lesson Study with enabling the implementation of new teaching approaches (Lewis 2002; Lewis and Tsuchida 1998b; Stigler and Hiebert 1999; Yoshida 1999b).

Although Lesson Study is commonly used by teachers and schools to improve teaching and learning in general, Lesson Study is also used to seek practical ideas for the effective implementation of the Course of Study (COS), the national curriculum (Murata and Takahashi 2002). This is a very common focus of school Lesson Study work during the transition period from one COS to a new COS.

During this transition stage, Japanese schools, especially public schools, typically conduct school-wide Lesson Study events for all the teachers at the school to work collaboratively to address the new curriculum implementation. The following case study shows how this process worked at one public elementary school (Takahashi 2014b).

4.4.1 The School Research Organization and the Research Steering Committee

The school the author examined in this section (Takahashi 2014b) is a public elementary school in Tokyo with about 760 students in grades 1 through 6, and 64 teachers and staff. Immediately after the Japanese Ministry of Education released a revision of COS, the teachers at the school decided to focus their Lesson Study work over the next two years on developing students' ability to express their ideas and learn from each other, which was a new point of emphasis in the revised COS.

During the two years of the school-wide Lesson Study, all full-time teachers at the school worked within a structure based on existing grade-level groups (see Fig. 4.2).

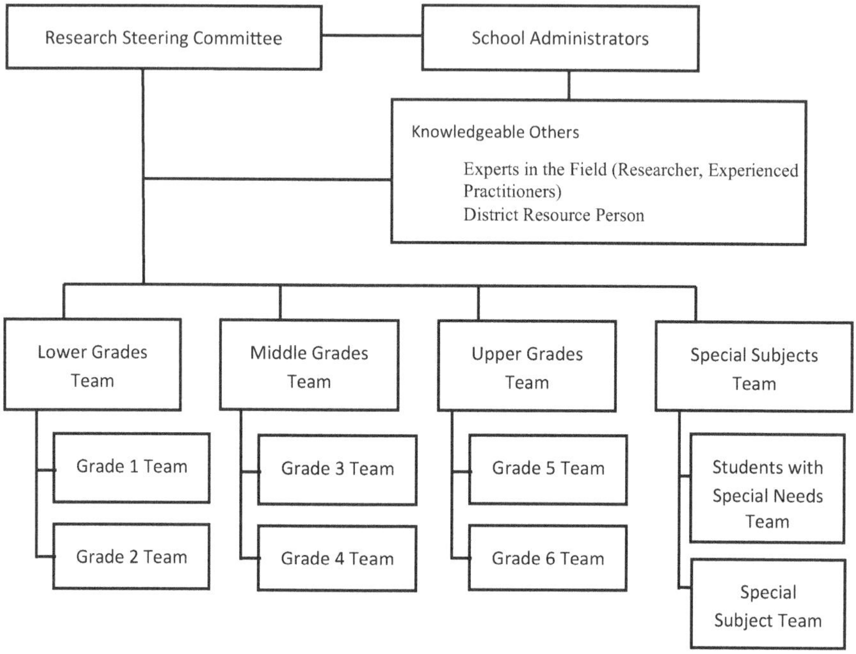

Fig. 4.2 Structure of the school research organization (reprinted from Takahashi 2014b with permission of Springer)

Grade-level groups typically exist in Japanese elementary schools to facilitate the sharing of responsibilities for running school events and for academic activities. Most public schools have time for grade-level meetings in their weekly schedule, typically about one hour. Teachers have desks in a common work area so that they can collaborate on a regular basis. In order to conduct the school-wide lesson effectively, each grade-level group was made responsible for crafting a plan for a research lesson, conducting their research lesson in front of the rest of the faculty, serving as panelists during the post-lesson discussion, and supporting the other teams' research lessons. The school also had grade-band teams, which consisted of all the teachers from adjacent grades, such as grade 1 and grade 2. Although the responsibility for lesson planning belonged to each grade group, most of the lesson planning was done in grade-band meetings in order to maintain consistencies across the grades and to help the teachers develop a shared view of the scope and sequence of the curriculum in adjacent grades. Finally, the grade-band meetings provided more opportunities for each teacher to participate in research lesson planning, a valuable experience especially for novice teachers not only to learn how to design lessons but also to deepen their understanding of the topics they teach.

Following common practice, the school organized a research steering committee, which consisted of representatives of each grade level and the lead teacher for mathematics,[1] who was appointed chairperson of the committee by the principal on the basis of his leadership ability and knowledge of mathematics teaching and learning. The committee led the school's efforts and maintained the cohesiveness of ideas across the grades. Among other things, the research steering committee was responsible for the following:

- Develop a master plan for the school research.
- Schedule and lead monthly meetings to find strategies to address the school's research theme based on the ideas of the teachers.
- Publish a monthly internal newsletter to record the findings from each research lesson.
- Plan, edit, and publish the school research reports, including those for the research open house.
- Arrange for knowledgeable others to present lectures, teach demonstration lessons, and give final comments at research lessons.

The first task of the research steering committee was to propose a focus for the school's research. That proposal was discussed by the entire faculty at the first faculty meeting of the school year. The following was the approved research theme and focus of study:

Research theme:	The development of individual thinking and the expression of these thoughts.
Focus of study:	Seeking effective ways to support students' individual problem solving skills and better facilitation of whole-class discussion in teaching through problem solving.

The research theme articulated a goal for students while the focus of study expressed the faculty's idea about a path toward accomplishing the goal.

Each grade-level team developed a lesson plan for a research lesson and conducted the research lesson and post-lesson discussion to address the theme. Most of the research lessons were scheduled on one of the half-day professional development days in order for all full-time teachers to be able to observe the lessons and participate in the discussions. As a result, each full-time teacher had the opportunity to be a part of eight research lessons during one school year. The school also invited two distinguished mathematics educators to give lectures, one professional development day in the first month of the school year (April) and another during the summer break, about the issues and trends in mathematics education and ideas for implementing the new COS.

Throughout the two years of the project, the research steering committee met between the research lessons to summarize the ideas that had been proposed by each lesson planning team and addressed during the post-lesson discussion. They published their summaries as a school research newsletter each month. These

[1]The lead teacher has his or her own self-contained class but also has responsibility for providing support for the upper grade teachers and for preparing curriculum materials for the school.

newsletters documented the process of this long-term collaborative effort, and, more important, they allowed the teachers to share what was discussed and helped other teams build off the results of previous research lessons.

4.4.2 Lesson Plans and Their Development

In each stage of lesson plan development, members of the research steering committee reviewed the lesson plan and provided feedback to the team. Through this process, they tried to ensure that all the lesson plans developed by the school were of sufficient quality to contribute to the school's effective implementation of the new COS. In order to do so, the committee distributed to each teacher the following list of questions to guide them toward higher quality lesson plans:

- Does the lesson plan provide sufficient information for the teacher to understand the task and the flow of the lesson?
- Does the lesson plan provide sufficient information about how the planning team decided to teach the lesson as described by the plan?
- Do the objectives of the lesson plan clearly address the Course of Study?
- Are the tasks appropriate for the students given the date of the lesson?
- Are the key questions clear? Will they encourage students to think mathematically and help them complete the task independently?
- Does the lesson plan include reasonable anticipated student responses and indicate how the teacher will help students overcome any misunderstandings?
- Does the lesson plan include a plan for formative assessment and a plan to accommodate individual student differences during the lesson?

4.4.3 Disseminating the Results of the School Research

Toward the end of Year 2, the school faculty and staff hosted a half-day public open house to share their findings. All content specialists of the district and principals of other area schools were invited to the open house, and many other schools sent their teachers. In all, a total of 612 participants, including teachers, administrators, educators, and parents attended this event.

The public open house consisted of three major parts: public research lessons, research presentations by the school's research steering committee, and a panel discussion by experts in the field of mathematics education who had been involved with the school's research project. There were 28 mathematics lessons conducted simultaneously based on 25 different lesson plans available for the participants to observe at the beginning of the open house. All 25 lesson plans were in a booklet given to each participant on arrival at the school. The participants were able to witness strategies for the effective implementation of the COS in live lessons and were able to bring these ideas back to their own school as a set of lesson plans. The presentation given by the members of the steering committee informed the participants about the philosophy and the rationale behind the strategies being used at the school. The presentation also provided educators from other schools an opportunity to learn how the school conducted its research using Lesson Study and what the faculty at the school had learned by going through this process.

Two sets of research reports were also made available for teachers and administrators of other schools as summaries of the school research effort of Year 1 and of Year 2. Since the school used a district grant to produce them, all the research reports were made available for free. In the second year, the school compiled a report covering the entire two-year study. The report was produced as four booklets: three of them were distributed at the public open house and the last was sent to all the schools in the district at the end of the school year. An English translation of one of these booklets is available at http://www.impuls-tgu.org/en/resource/readings/page-26.html.

4.4.4 The Results from the School-Wide Lesson Study

The Japanese national standards released in 2008 contained a new emphasis on having students learn to express their ideas and learn from each other, as a way to help students develop their own thinking. The teachers at this school chose to spend two years working through Lesson Study to research changes in practice that would address this new emphasis. Some of what they learned—and what they put into practice—is evident in the booklet they published for the open house. Here are a few points from the booklet

- Students were able to express their ideas using not only words but also mathematical expressions and diagrams. Because of the cohesive use of diagrams, such as tape diagrams, area diagrams, number line diagrams, and of expressions and equations throughout the grades, whole-class discussions became deeper and productive. Moreover, students were able to express their ideas in similar ways regardless of who was teaching the lessons.
- By crystallizing what was expected of students in each stage of problem solving (e.g., understanding the problem, solving the problem, reflecting upon the solution) and at the major points of teacher instruction, students were able to learn independently.
- By preparing effective key questions for each stage of problem solving, students were able to express their ideas in various ways and to talk to each other clearly by focusing on what should be discussed.
- By planning blackboard writing, the flow of the lessons became more coherent. Students became able to look back at what they learned by looking at the board. Then they could use it to put the various ideas together in integrated and expanded ways, and to evaluate their learning during the lessons by themselves.

Each teacher was deeply involved in planning only one research lesson per year, which may not seem like enough to support such profound growth. But the school's work over the two years was carefully organized to support teacher learning in various ways. Each teacher at the school had at least two opportunities to critique lesson plans from another team during the planning process through the grade-band meetings. Teachers observed and discussed the lessons of all the other grades at the school. And the newsletters published by the research steering committee helped each successive team build on what was learned before.

4.4.5 Supporting the School-Wide Lesson Study

When implementing new ideas of teaching and learning, teachers must figure out what the necessary changes will look like in their own classrooms and with their own students. To do so, teachers need to conduct their own research, and Lesson Study provides an organized way to do so. Since Lesson Study is tied directly to teachers' practice, teachers can minimize the gap between research and practice. Outside of Japan, many Lesson Study projects have been conducted by a few volunteers within a school or across school districts. Individual teachers can certainly improve their own teaching by participating in such volunteer groups. But in Japan, as this case study illustrates, improving teaching is a responsibility of all teachers at a school, to be worked on together.

Although teachers work hard to improve teaching and learning by collaborating with their colleagues through Lesson Study, they can be limited by what they do not know. In order to maximize the effect of the collaboration, Japanese school administrators usually provide additional supports for expanding teachers' knowledge. These include a structure to support collaboration (grade-level teams and grade-band teams) and distributing leadership in the form of a research steering committee that comprises teachers from multiple grades, and access to new knowledge and expertise via an outside expert.

Researchers have noted the importance of outside expertise provided by the so-called "knowledgeable other" in making Lesson Study effective (Lewis et al. 2006; Lewis and Tsuchida 1998b; Takahashi and Yoshida 2004; Yoshida 1999b). The following section describes the roles of the knowledgeable other in school-wide Lesson Study.

4.4.6 The Role of the Knowledgeable Other

It is common practice among Japanese schools to bring in an outside expert who is knowledgeable about the school research theme. This person is referred to as a "knowledgeable other." Based on a study conducted by the Takahashi (2014a), the knowledgeable other is responsible for

(1) bringing new knowledge from research and the standards,
(2) showing the connection between the theory and the practice, and
(3) helping others learn how to reflect on teaching and learning.

Each of the responsibilities is elaborated below.

4.5 Bringing New Knowledge from Research and the Standards

As in many other countries, most Japanese elementary school classroom teachers have to teach all subjects. In order to update their knowledge regarding mathematics teaching and learning, teachers need support from people who have access to the latest research and the standards. One of the important responsibilities of a knowledgeable other is to help classroom teachers deepen their understanding of the content, the curriculum, ideas behind the textbooks, and pedagogical ideas.

When teachers engage in lesson study, they are expected to deepen their knowledge of mathematics teaching and learning by reading teacher resources such as teaching guides, recent journal articles, and curriculum materials, as well as carefully studying the textbooks that the school uses. But Japanese educators often emphasize that simply reading about what research and the standards say is not enough.

Like students, teachers best learn new ideas with concrete examples. A research lesson serves as a rich source of concrete examples from which teachers can learn, if given proper guidance. Thus, Japanese schools customarily invite a knowledgeable other to their research lesson and ask that person to provide "final comments," lasting 10, 30 min, or more at the end of the post-lesson discussion.

In the final comments, knowledgeable others typically begin their comments by providing new knowledge pertaining to the teaching and learning of the topic of the lesson, drawing from the COS and from the textbooks. They then examine key ideas in the lesson plan. Finally, they reflect on the actual events of the lesson, bringing up specific evidence of what students had learned from the lesson, and make suggestions for future consideration.

In order to do so, the knowledgeable others may prepare in advance some handouts for the teachers based on the draft lesson plan they receive a week before. These handouts mainly elaborate on that part of the COS related the topic of the lesson so that the teachers other than the lesson planning team can understand the fundamental ideas involved. Although a similar curriculum investigation is often done by the lesson planning team as a part of their lesson planning research, the handout aims to go beyond the investigation by the team. This is the first place where a depth of understanding of the contents and the curriculum is required of the knowledgeable other.

Another important responsibility of the knowledgeable others is to elaborate on the ideas behind the textbook pages. Japanese textbooks are thin but contain rich content with a focused and coherent organization (Watanabe et al. 2010). The underlying rationale for textbook content is sometimes subtle, and it can be difficult for teachers to see the connections between the problems on different pages, in different units, and from different grades. The knowledgeable other can help teachers see those connections.

All of the above aims to deepen the teachers' knowledge and understanding of the contents of the research lesson.

4.5.1 Showing the Connection Between Theory and Practice

Japanese teachers understand that they are responsible for implementing the COS, using theory and research findings to improve students' attainment of the curriculum. Although there are plenty of materials available for teachers to help them implement the curriculum, reading these resources or listening to experts' lectures are usually not enough to develop the expertise to use them in Level 3 teaching. In Lesson Study, teachers have the opportunity to plan lessons based on knowledge acquired by reading or listening, to teach the lesson based on a carefully designed lesson plan, and to reflect on the teaching and learning using evidence from the lesson. Through this process, teachers can try out new ideas or practices and evaluate the effectiveness of their lesson plan in the post-lesson discussion.

School-wide lesson study almost always focuses on a research theme selected by the faculty. Therefore, to help the school translate theory into practice, another important role of the knowledgeable other is to connect what they observed during the research lesson to the school's research theme. The knowledgeable other should try to highlight concrete evidence from the lesson that is relevant for assessing the progress of the school toward its research theme.

Knowledgeable others may also suggest possible directions the school should take in order to pursue the research theme, and may also offer professional viewpoints and opinions about the school research and the research lessons.

4.5.1.1 Helping Others Learn How to Reflect on Teaching and Learning

Another role of the knowledgeable other is to help the school conduct effective post-lesson discussions. Thus, he or she should give the teachers the opportunity to reflect upon important lessons learned from the discussion and on what else they could learn if the discussion were improved. In order to do so, the knowledgeable other should be able to not only summarize the discussion, but also to effectively contribute to the discussion by raising important issues that were not addressed during the post-lesson discussion.

Sharing what the knowledgeable other observed during the lesson helps the teachers see what can they learn if they have good "eyes for observing students."

In fact, knowledgeable others often say that observing lessons with experienced Lesson Study practitioners was the best way to develop good eyes for observing students.

4.6 Recommendations

Stigler and Hiebert (1999) argue that Japanese mathematics lessons better exemplify recent reform ideas than do US lessons. One of the reasons Japanese teachers are able to use reform ideas effectively in their classroom is their participation in Lesson Study. Lesson Study provides the opportunity for classroom teachers to work collaboratively to seek effective implementation of new ideas, rather than struggle in isolation to understand how the ideas look in his/her own classroom.

Since the early research on Lesson Study published late 1980s (e.g., Lewis and Tsuchida 1998a; Stigler and Hiebert 1999; Yoshida 1999b), researchers, educators, and teachers around the world have attempted to replicate its success at transforming Level 1 and 2 teaching to Level 3 teaching focused on problem solving (e.g., Hart et al. 2011). Although many schools and teachers have tried to use ideas from Lesson Study in various ways, only a few cases have been documented in which there was strong evidence of impact on teaching and learning (e.g., Gersten et al. 2014; Lewis et al. 2006).

One of the reasons behind is that the "Lesson Study" outside Japan based on early research documents may have left some important aspects of Lesson Study in Japan. For example Fujii (2014) argues that many activities described as "Lesson Study" are often very different from Lesson Study in Japan.

In order to overcome such dilemma and seek vital impacts on student and teacher learning, several new projects with careful examination of critical aspects of Lesson Study in Japan to design a comprehensive program to support teachers and schools have been conducted in the US, UK, and some other counties. Although full reports of these projects may not be published, some important aspects of Lesson Study may be revealed (e.g., Takahashi and McDougal 2016).

For educators who try to improve mathematics teaching and learning, it is important to understand why Lesson Study has been less consistently impactful outside of Japan and design a program carefully so that the teachers can receive appropriate support in order to experience an authentic Lesson Study process.

References

Ball, D. L., Thames, M. H., & Phelps, G. (2008). Content knowledge for teaching: What makes it special? *Journal of Teacher Education, 59*(5), 389–407. doi:10.1177/0022487108324554

Fujii, T. (2014). Implementing japanese lesson study in foreign countries: Misconceptions revealed. *Mathematics Teacher Education and Development, 16*(1), 65–83.

Gersten, R., Taylor, M. J., Keys, T. D., Rolfhus, E., & Newman-Gonchar, R. (2014). *Summary of research on the effectiveness of math professional development approaches. (REL 2014–010).* Retrieved from Washington, DC: http://ies.ed.gov/ncee/edlabs

Hart, L. C., Alston, A., & Murata, A. (Eds.). (2011). *Lesson study research and practice in mathematics education.* Now York: Springer.

Lewis, C. (2000, April 2000). *Lesson study: The core of Japanese professional development.* Paper presented at the AERA annual meeting.

Lewis, C. (2002). *Lesson study: A handbook of teacher-led instructional change.* Philadelphia: Research for Better Schools, Inc.

Lewis, C., & Tsuchida, I. (1998a). A lesson is like a swiftly flowing river: How research lessons improve Japanese education. *American Educator, 22*(4), 12–17, 50–52.

Lewis, C., & Tsuchida, I. (1998b). A lesson like a swiftly flowing river: Research lessons and the improvement of Japanese education. *American Educator, 22*(4).

Lewis, C., Perry, R., Hurd, J., & O'Connell, M. P. (2006). Lesson study comes of age in North America. *Phi Delta Kappan, 88*(04), 273–281.

Makinae, N. (2010). *The origin of lesson study in Japan.* Paper presented at the The 5th East Asia Regional Conference on Mathematics Education: In Search of Excellence in Mathematics Education, Tokyo.

Murata, A., & Takahashi, A. (2002). Vehicle to connect theory, research, and practice: How teacher thinking changes in district-level lesson study in Japan. In *Proceedings of the twenty-fourth annual meeting of North American chapter of the international group of the Psychology of Mathematics Education* (pp. 1879–1888).

National Mathematics Advisory Panel. (2008). *Foundations for success: The final report of the National Mathematics Advisory Panel.* Retrieved from Washington, DC.

National Research Council. (1989). *Everybody counts: A report to the nation on the future of mathematics education.* Washington, D. C.: National Academy Press.

Polya, G. (1945). *How to solve it: A new aspect of mathematical method.* Princeton, New Jersey: Princeton University Press.

Stigler, J. W., & Hiebert, J. (1999). *The teaching gap: Best ideas from the world's teachers for improving education in the classroom.* New York: Free Press.

Sugiyama, Y. (2008). *Introduction to elementary mathematics education.* Tokyo: Toyokan publishing Co.

Takahashi, A. (2000). Current trends and issues in lesson study in Japan and the United States. *Journal of Japan Society of Mathematical Education, 82*(12), 15–21.

Takahashi, A. (2011). The Japanese approach to developing expertise in using the textbook to teach mathematics rather than teaching the textbook. In Y. Li & G. Kaiser (Eds.), *Expertise in mathematics instruction: An international perspective* (pp. 197–219). New York: Springer.

Takahashi, A. (2014a). The role of the knowledgeable other in lesson study: Examining the final comments of experienced lesson study practitioners. *Mathematics Teacher Education and Development, 16*(1), 4–21.

Takahashi, A. (2014b). Supporting the effective implementation of a new mathematics curriculum: A case study of school-based lesson study at a Japanese public elementary school. In I. Y. Li & G. Lappan (Eds.), *Mathematics curriculum in school education* (pp. 417–441). New York: Springer.

Takahashi, A., & McDougal, T. (2016). Collaborative lesson research: maximizing the impact of lesson study. *ZDM*, 1–14. doi:10.1007/s11858-015-0752-x

Takahashi, A., & Yoshida, M. (2004). How can we start lesson study? Ideas for establishing lesson study communities. *Teaching Children Mathematics, 10*(9), 436–443.

Watanabe, T., Takahashi, A., & Yoshida, M. (2008). Kyozaikenkyu: A critical step for conducting effective lesson study and beyond. In F. Arbaugh & P. M. Taylor (Eds.), *Inquiry into Mathematics Teacher Education, Association of Mathematics Teacher Educators (AMTE) Monograph Series* (Vol. 5, pp. 131–142).

Watanabe, T., Takahashi, A., & Yoshida, M. (2010). Supporting focused and cohesive curricula through visual representations: An example from Japanese textbooks. In B. Reys, R. Reys, & R. Rubenstein (Eds.), *2010 Yearbook: Contemporary issues in mathematics curriculum* (pp. 131–143). Reston, VA: National Council of Teachers of Mathematics.

Yoshida, M. (1999a, April). *Lesson Study [jugyokenkyu] in elementary school mathematics in Japan: A case study.* Paper presented at the American Educational Research Association Annual Meeting, Montreal, Canada.

Yoshida, M. (1999b). *Lesson study: A case study of a Japanese approach to improving instruction through school-based teacher development.* (Dissertation), University of Chicago, Chicago.

Chapter 5
Towards Balancing Knowledge and Practice of In-Service Mathematics Teacher Education Program in Korea

Oh Nam Kwon, Jung Sook Park, Jaehee Park and Jee Hyun Park

Abstract This chapter describes the general system and the changes of in-service mathematics teacher education in Korea. Korea is achieving excellent results in international comparative assessments such as PISA and TIMSS, and one of the reasons for this is the excellence of its teachers. Teachers' professional development is divided into two areas, preservice teacher education and in-service teacher education. This chapter focuses on in-service teacher training programs in Korea, analyzes the changes in the training contents and methods with regard to mathematics teacher training, and then discusses what goals should be set by Korean mathematics teacher training programs for enhancing professional development.

keywords In-service teacher training program · Teachers' professional development · Teacher education

5.1 Introduction

Teacher training for in-service teachers has been regarded as an important mechanism that enhances teachers' understanding of content knowledge and teaching method as well as the qualitative level of teaching practice in the classroom

O.N. Kwon (✉)
Seoul National University, Seoul, South Korea
e-mail: onkwon@snu.ac.kr

J.S. Park
Yangjae High School, Hanyang University, Seoul, South Korea
e-mail: pjs1924@sen.go.kr

J. Park
Gyeonggi Science High School, Suwon, South Korea
e-mail: udmji@snu.ac.kr

J.H. Park
Banpo High School, Seoul, South Korea
e-mail: jeannei@sen.go.kr

B. Kaur et al. (eds.), *Professional Development of Mathematics Teachers*, Mathematics Education – An Asian Perspective,
DOI 10.1007/978-981-10-2598-3_5

(Ball and Cohen 1999; Little 1993). Thus, teacher training is essential for improving the quality of education, and providing effective training is recognized as a starting point for enhancing the quality of education.

To improve the quality of education and introduce a new paradigm to schools, a change in the current paradigm of the teacher training system that affects teachers' professional development is needed. Many previous teacher professional enhancement development programs were based on the model of technological reasoning, which defines the roles and behavioral characteristics of teachers. In other words, it was expected that the quality of teaching would improve if teachers learned about the principles and procedures established through research (Doyle 1991). However, many studies show that simple transfer of knowledge is not effective in changing teachers' beliefs, attitudes, or teaching practices (Greenberg and Baron 2000).

Teaching, which is a teacher's duty, is practical like the work of doctors and lawyers. To carry out such practical work well and with professionalism, teachers need to acquire and improve practical knowledge combined with theory and practice (Cochran-Smith and Lytle 1999). Enhancing teachers' professional abilities does not start from only a deep understanding of content (Hill et al. 2005) and is not possible through instruction given directly by external experts. Rather, the methods used to enhance teachers' professionalism should be based on authentic contexts that reflect teachers' direct participation in the definition and formation of the difficulty of practice in an actual classroom situation or a context teachers' encounter (Cobb et al. 2003; Kazemi and Franke 2004). Today, teacher training programs are carried out in connection with the context of teaching practice and lead to the development of class assignment or activities, on the basis of research findings that teachers' professional development is achieved in close relationship with the context of their teaching practice (Cochran-Smith and Lytle 1999; Shulman and Shulman 2004).

Teacher training in Korea is also changing from theory-centered to practice-centered training. Korea is achieving excellent results in international comparative assessments such as Programme for International Student Assessment (PISA) and Trend in International Mathematics and Science Study (TIMSS), and one of the reasons for this is the excellence of its teachers. In fact, because excellent preservice teachers become teachers in Korea, their efforts in developing their own capabilities, such as spending considerable time for self-development, are large on average (Kwon and Ju 2012). An example of this is that the number of Korean teachers with a Master's degree and/or doctoral degree is on the rise, as teachers attend graduate school with a personal goal. Although 17.4 % of elementary school teachers had a Master's degree and/or doctoral degree in 2005, this number was 26.9 % in 2012, representing an increase of 9.5 % points in 8 years. This phenomenon is found in middle school teachers as well. In 2012, 36.8 % of all middle school teachers and 39.9 % of high school teachers had a Master's and/or doctoral degree. Generally, there is a phenomenon that the educational level of teachers increases as the school level increases (Korean Educational Development Institute 2013).

In Korea, in-service teachers must complete at least 90 h of professional development activities to upgrade their teaching certificate (usually after 3–4 years of teaching). Following which, they are required to participate in professional development activities every year (Sami 2013). In 1995, the New Educational Reform Plan was announced and it states that in-service teachers must receive training at regular intervals. However, the plan did not provide any mandatory details about the number of hours and regularity of the intervals, such as every year or three years (Education Reform Commission 1996). In recent years, the government fully supports a minimum of 20 h of annual professional development for each teacher. However, most teachers attend 40–60 h of professional development activities to keep up with the new developments in their fields of expertise.

The 15th ICMI Study (Even and Ball 2010) divides teachers' professional development into two areas, preservice teacher education and in-service teacher education. As the research on preservice teacher education in Korea was described in Kwon and Ju (2012), this paper explores in-service teacher training programs in Korea, analyzes the changes in the training contents and methods with regard to mathematics teacher training, and then discusses what goals should be set by Korean mathematics teacher training programs for enhancing professional development.

5.2 Teacher Training Programs in Korea

Immediately after liberation, the teacher training programs in Korea had been carried out sporadically without a legal basis or separate training institutes. From the time the law on teacher training institutes was promulgated and amended in 1953, 1964, and 1972 to the present, such institutes have been the foundation of in-service teachers' training programs (Lee et al. 1993). As "Provisions for teacher training programs" was enacted by a presidential decree in the late 1980s, a system was organized for the type and establishment of training institutions, training targets, training types and courses, training period, and records of training performance (Shin and Jeon 2008). Authorized distance educational training institutes, which began operations in 2000, have continuously expanded; there were 39 distance educational training institutes in the second half of 2001. Today, there are over 70 distance educational training institutes in operation. Given the advantage that one can receive training in a variety of forms anytime and anywhere, teachers' demand for and interest in distance educational training programs are continuously increasing (Kim and Kim 2013).

Despite the continuous effort of the government to improve the teacher training system, many studies pointed out the problems of the teacher training system in the early 2000s (Shin and Jeon 2008). Representative problems were that the training programs did not reflect the demands of the fields, leaned toward the use of a one-sided lecture-style training method, relied excessively on institute-centered training, and were lacking in variety. In response, in 2001, the Ministry of Education started to systematically evaluate the curriculum and the operation status of training programs by introducing "Teacher training certification for the evaluation of training institutes." Also, by designating various excellent teachers' associations as training institutes for special fields, it jointly developed various training programs and materials and implemented practice-oriented training programs directly related to teaching. Also, it attempted to foster and support voluntary training programs or research activities by allowing the designation of school-based training programs or autonomous training programs of nearby school associations as training programs for special fields.

Keeping pace with the current trends, since 2000, teacher training programs have been diversifying in terms of their contents and methodological aspects. Moving away from institute-centered training programs, school-based, or individual-centered training programs are growing, and institute-centered training programs are also diversifying due to the training programs run by the Ministry of Education or the Office of Education as well as various teacher groups or associations. Particularly, as the voice for applicability on-site increases, school site-centered training programs are being developed, moving away from one-sided, lecture-style training method (e.g., Kwon et al. 2014). Also, through "Plans to Enhance the Teaching Professionalism of Teachers," the Ministry of Education (2009) is focused on supporting teachers' professional development through open classes, teaching consulting, and teaching clinics. In addition, it required teachers to continuously seek to improve their teaching practices by enacting the law on "Evaluation for teacher's ability development" in 2011, which makes training programs mandatory when a teacher fails to receive a certain level of assessment from students.

As shown in Fig. 5.1, teacher training programs are currently divided into training institute-centered training programs, school-based training programs, and individual-centered training programs (Ministry of Education 2013).

Institute-centered training programs are divided into qualification training, job training, and special training as the training programs led by various training institutes. Training programs for obtaining higher qualifications (e.g., level 1 or 2

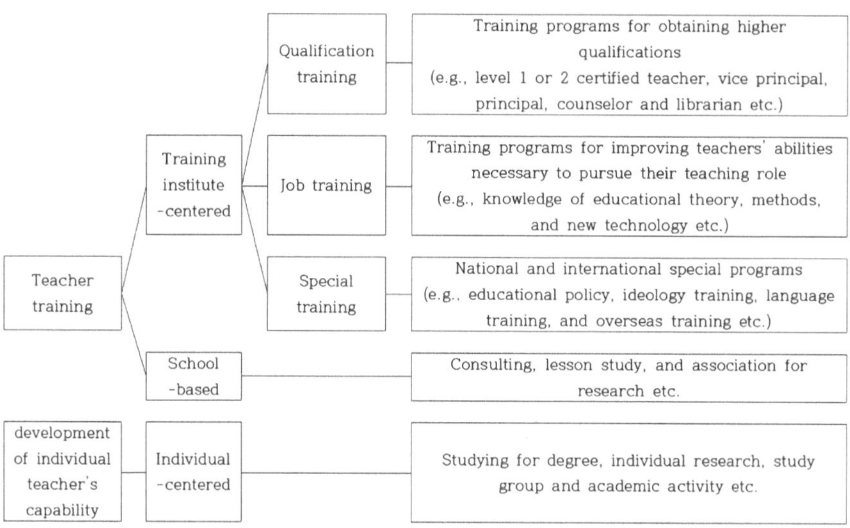

Fig. 5.1 Types of teacher training programs (Ministry of Education 2013, p. 2)

certified teacher, vice principal, and principal) and training programs for obtaining special qualifications (e.g., counselor or librarian) belong to qualification training programs, which are training programs for obtaining qualifications. Job training programs are implemented for teachers to improve their knowledge of educational theory, methods, and general education; learn new technology; and cultivate the skills and qualifications necessary to pursue their teaching role. Special training programs are training programs in which state or local government requires teachers to complete national and international special programs such as educational policy, ideology training, language training, and overseas training of inspection tour for cultivating professional knowledge and skills (Ministry of Education 2013).

In 2014, there were 167 institutes authorized or established to implement the institute-centered training programs. Seventeen city and provincial education training institutes, eight educational administration, and general education institutes, 80 education training institutes affiliated with universities, and 62 distance education training institutes are leading in ensuring the substantiality of the qualification training programs and the job training programs for teachers' professional development.

Mathematics teachers in Korea complete a mandatory new teacher training program when they are hired and mandatory qualification training program for level 1 certified teacher after about 5 years of employment. They complete about 60 h of job training every year. Concrete changes in the contents and methods of training programs for mathematics teachers are discussed in Sect. 5.3.

5.3 Mathematics Teacher Training Programs in Korea

The contents of training programs for mathematics teachers in Korea are changing in accordance with the changes in the curriculum and government policies. Similarly, training methods are changing with the current trends.

5.3.1 Changes in the Contents of Training Programs

Since 1993, qualification training programs have required trainees to complete 18–36 h (10–20 %) of liberal arts courses, 18–36 h (10–20 %) of teaching courses, and 126–144 h (60–80 %) of major courses in order to reduce the regional gap in training contents (Lee et al. 1993). To specifically explore how the contents of the training program for level 1 certified teachers are changing in accordance with the current trends, the training contents for level 1 certified teachers provided by the Seoul Education Training Institute are compared and shown in Table 5.1. Table 5.1 compares hours in accordance with the content area of the training program for level 1 certified teacher implemented in 1993 and 2014.

Table 5.1 Comparison of the 1993 and 2014 curricula of the training program for level 1 certified teacher

1993 training program for level 1 certified teachers				2014 training program for level 1 certified teachers			
Liberal art subject	Pedagogy course	Major course	Total	Basic literacy	Capability area	Major area	Total
20 h (11 %)	36 h (20 %)	124 h (69 %)	180 h	15 h (16 %)	22 h (24 %)	54 h (59 %)	91 h

When the numbers in Table 5.1 are compared, the hours required for completion are cut in half from 1993 to 2014, and the proportion of the major course is also slightly lowered. Since 2011, training hours were adjusted from over 180 h to over 90 h. In 1993, there were five liberal art subjects, including "Understanding the cultural tradition," "Information society and computers," and "Environment and education," and nine pedagogy courses, including "Problems with youth and guidance," "Trends of educational reform," and "Educational law." Among the nine major courses, 12 additional courses were related to the mathematics content knowledge. Examples of specific courses include "Sequence and limit," "Equation and inequality," and "Calculus." Courses related to pedagogical content knowledge include "History of mathematics," "Mathematics education and assessment," and "Mathematical problem solving." Regarding such contents of the training programs, Jung et al. (1994) pointed out the following problems: Most courses in the qualification training programs fail to have relevance with secondary school mathematics, training is not designed from an integrated perspective on mathematics or with emphasis on the application of mathematics, and most courses are carried out in the form of expository instruction.

There were many negative views of the qualification training program by teachers. The level 1 certified teacher training program is a representative institute-centered training program that most teachers have to take. Kim and Kim (2005) stated that teachers add the meanings "rite of passage" and "picking score" to the course of the level 1 certified teacher training program. This indicates that the programs are regarded as courses that all teachers should take and that, although trainees are qualified as level 1 certified teachers when they fulfill the basic required hours and grades, they are aware that they should earn high grades, as the grades affect their promotion.

In the qualification training program implemented in 2014, the titles of the courses offered in the program have changed from liberal art subject to basic literacy, pedagogy course to capability area, and major courses to major area. There were seven basic literacy courses, including "Global education policy trends," "Use of smart work beyond the classroom," and "Living together and understanding special education," and 10 courses belonging to the capability area, including "Understanding and practices of a happy class," "Use of materials for good classes," and "Understanding student counseling." There were 12 courses belonging to the major areas. Six courses related to the mathematical content knowledge included "Secondary geometry," "Secondary algebra," and "Secondary probability and statistics", other six courses related to pedagogical content knowledge included, such as "Storytelling and mathematics education," "Mathematics, personality, and mathematics education," "Math class observation and analysis," and "Math classes with GeoGebra."

The 6th curriculum was applied in 1993, and the 2009 revised curriculum was applied in 2014. This indicates that there were many changes in the courses in accordance with the changes in the curriculum. Given that "problem-solving capability," which is the focus of the 6th curriculum, was established for major courses and "creativity and personality," which is the core of the 2009 revised curriculum, was established as "Mathematics, personality, and mathematics education," one of the goals of the teacher training programs is to widely spread the national education policy. Based on the subject names, it seems the courses of the teacher training programs implemented in 1993 are focused on content knowledge, whereas the teacher training programs implemented in 2014 are focused on pedagogical content knowledge. The previous paradigm of the teacher training programs was to train teachers to have necessary knowledge, but recent teacher training programs are more interested in class and students. Also the paradigm of the recent teacher training programs is focused on participation and practice. In particular, in the course "Class observation and analysis," trainees have an opportunity to reflect on teaching by directly filming their own teaching and analyzing it.

The results of Park and Moon's (2009) survey showed that 39.5 % (less than half) of mathematics teachers did not participate in a mathematics training program within 10 years or completed less than 5 h of training. The first reason is the lack of established mathematical training programs, and the second reason is the use of contents that are not applicable on-site. Teachers want training programs that provide practical knowledge and methods that they can actually use in the classroom. In fact, as shown in the analysis of the status of mathematics teacher training programs by Lee and Jang (2012), the training contents that were evaluated as "well done" by the Office of Education were those that can be used in an actual classroom site. Given that the goal of professional development was to "help teachers develop instructional practices in which they induct their students into the ways of reasoning of the discipline by building systematically on their current mathematical activity" (Cobb and McClain 2001, p. 207), it is desirable to modify the qualification training program to include practical knowledge that can be used for a class.

One of the reasons that the contents of the qualification training program changed in such a way is that the teachers themselves formed groups for self-development because the institute-centered training programs failed to reflect their demands. A representative voluntary meeting of mathematics teachers is the Korean Society of Teachers of Mathematics. The Korean Society of Teachers of Mathematics started with the publication of a magazine called *Math Love* in 1995. Currently, it is a research society of mathematics teachers with the goal of popularizing mathematics. It is run purely on membership fees, participation, and donations from members. By operating a seminar team, it studies teaching-learning methods, philosophy of mathematical education, curricula, the use of software in mathematics classes, and the history of mathematics. Also, it develops teaching materials that are directly incorporated in teaching. It regularly holds a conference in which the seminar team shares its research findings. The Math Festival, which began in 1999, has been promoted as a part of the job training programs since 2004 and is one of the representative jobs training programs for which in-service mathematics teachers voluntarily apply. Table 5.2 summarizes the training contents of the Math Festival in 1999 and 2013 (http://www.tmath.or.kr/bbsd/list.asp?bbsId=data_30_16).

Lecture contents in the first Math Festival in 1999 included content promoting changes in mathematics teaching like "Mathematics class introducing the history of mathematics" as well as class organization methods and evaluations, like "Differentiated class," and mathematics teaching that uses technology. When compared to the level 1 certified teachers training program implemented in the same

Table 5.2 Training contents of the Math Festival in 1999 and 2013

	Content area	Lecture title
1999	"A" session	Math lessons with activities, Practice of differentiated class, Mathematics education introducing the history of mathematics, Let's make mathematics class, …
	"B" session	Secondary school performance assessment, Differentiated class, Potential use of Excel in statistics class, Use of GSP in secondary school geometry education, …
	"C" session	Tessellation, Probability in everyday life, How to use GSP in class, Fractal geometry that is realized in the classroom by using a graphing calculator, …
2013	Plenary lecture	One gets creative when one sees nature through the eyes of mathematics, King Sejong's gunpowder weapons and precision science and technology
	"Study of teaching materials" session	Probability of meeting n people at the same time and problem of shortest distance, Between the infinite and the limit, Mathematics class in which all can participate, …
	"Improving teaching" session	Bridging art and mathematics, Mathematics class that nurtures convergence talent based on creativity and personality, Educational innovation in a learning community, Hanwool Middle School, …
	Math education session	Needs of ecological mathematics and practice, Mathematical thinking and creative thinking, Storytelling-based strategy for mathematics class, …
	Experience/gifted mathematics session	Travel story with mathematics, Practice of running an experimental mathematics program at a secondary school, Plan for vitalizing a math club, …
	Observing classroom session	Restructuring class to empower students, Why watch classes, Observing the middle school class on the properties of quadrilateral, Observing a high school class on the application of derivative, …
	Workshop	Elementary GeoGebra, Advanced GeoGebra, Board games and mathematics, Dual of a regular polyhedron, Making a math book, …

year, this forms a training program by focusing on pedagogical content knowledge, which is the knowledge teachers need to run a class. Furthermore, only three out of a total of 30 lectures were taught by professors; the other 27 instructors were in-service teachers. The Math Festival in 2013 was organized more systematically. The content was divided into two lectures and five sections of study on teaching materials, teaching improvement, mathematics education, experience/gifted mathematics, and observing classrooms, and there were seven lectures per section and 12 additional workshops. Accordingly, the festival presented information that was useful to teachers for improving their teaching and observing classrooms and collaborative classes or activity classes, therefore moving in the direction of greater emphasis on practical knowledge.

5.3.2 Changes in the Training Methods

In the past, teacher training programs have been largely lecture-oriented, and this is still true today. As the main agents of realization of the teacher-learning process, teachers have to determine the specific class situation of their school site and thus, even if they have attended a good lecture, the effect of the training program is not large if they do not actually apply the ideas in class (Park et al. 2010). In this context, the Ministry of Education (2013) is actively suggesting the introduction of various training methods such as workshops, case presentations, group discussions, and site visits in order to move beyond the lecture-oriented training program and maximize the training program effects.

Recent changes in teacher training methods are in accordance with the perspective that sees teaching as practical knowledge (Shulman and Shulman 2004). As a unique professional knowledge of teachers, practical knowledge, which is a concept defined by Elbaz (1981, 1983), refers to knowledge that is generalized and reconstructed in accordance with an encountered situation on the basis of values and beliefs. Training methods such as workshops, case presentations, group discussions, and site visits were suggested as methods that lead teachers to develop practical knowledge. However, these training methods also focus on the growth of individual teachers, and thus, a teacher community in which teachers focus on students and jointly reflect and discuss educational practice was recently formed.

Teacher community is a training method that emphasizes continuity, cooperation, and solidarity as an alternative that can overcome the limitation of the traditional approaches in which teacher professional development is focused on an individual (Kwon et al. 2014). Teacher community, which shares a similar school context, learning environment, familiarity among members, and objective of jointly agreed teaching activities, induces teachers' mutual development. In addition, teachers are regarded not only as people who runs the curriculum but also as practical researchers belonging to the teacher community.

By organizing a teacher community based on teachers within the same school, Kwon et al. (2014) developed a mathematics teacher training program that continuously fostered the development of other teachers, even after the training. What distinguished it from other teacher communities is that it targeted three teachers teaching the same grade at the same school as one team and allowed them to plan to teach together in the training process; also, continuity was maintained even after the training ended. Training content was about convergence teaching for secondary school mathematics teacher and storytelling mathematics for elementary school teacher which have become issues in today's curriculum. Teachers who participated in the training program gained confidence as they planned and applied a new teaching method along with basic theory and reflected on the needs of the community and continuous maintenance. The program offered them an opportunity to share and understand their knowledge of the teaching process. Thus, the teacher community can be a useful teacher training method that enhances individual professional development and growth through cooperation and reflection within the community.

Another training that expanded and changed teacher training in Korea is distance training, accompanying the revitalization of e-learning. Because distance training programs allows anyone to receive training anywhere and anytime using information and communications technology, they are effective for teachers who have difficulty securing time for training due to their teaching responsibilities and various administrative duties. In addition, as teachers directly experience education that uses information media, distance training has the advantage of playing a leading role in changing school education for the information age (Cho 2004; Jeong et al. 2009).

Distance training programs have continued to grow since their inception in December 2000, and as of 2013, 62 distance educational institutes were in operation. It was found that 67 % of those who participated in a distance training program favored it over collective education training programs (15 %) or blended training programs (16 %), and 95 % were willing to participate in distance training again. Although 1820 people received distance education training in 2000 when the distance training programs first began, 288,030 people completed distance training programs annually as of February 2010. Accordingly, distance training programs are growing rapidly, now accounting for about 40 % of all training programs (Korea Education and Research Information Service 2013). Moreover, although distance education training content was implemented only on web-based platforms, training content that enables one to learn on smart devices (smart phones and smart pads) has continuously developed due to the recent smart technology and widespread use of mobile devices. In this respect, it is expected that teachers' demand for and interest in distance education training programs that allow them to receive training in various forms anywhere and anytime will further increase. In 2009, the Korea Education and Research Information Service developed the Teacher Training Information Service (http://ttis.edunet.net) and are providing an information service that has the characteristics of a portal site for teacher training programs. Such cases are hard to find in other countries, and the program has significance as national comprehensive training service for teachers (Kim and Kim 2013).

However, criticism of distance training programs is also increasing. The training courses that were offered the most in distance education training institutes from January 2011 to September 2012 had the topics "self-development" (48.21 %), "teaching and learning practice" (21.38 %), and "life guidance and counseling" (16.37 %) (Kim and Kim 2013). As these three subjects accounted for 85.96 % of the whole, it was found that the training curriculum operated with too much emphasis on certain subjects. Particularly, there was few course related to mathematics. It was shown that the establishment of major courses is very insufficient, a teacher could not find the training he/she wanted. Furthermore, trainees are recognized as having listened to a lecture even if they did not listen to it properly (e.g., by fast forwarding through the video), and they can complete the training program simply by clicking. Thus, even though good lectures are provided, it may not be helpful to teachers. Also, the problem of securing excellent instructors has been pointed out. As a way to increase the efficiency of training, the training method of blended learning, which combines online training and offline training, is suggested, and training methods are expected to change continuously in the future.

OECD (2009) chose informal conversation that improves teaching as the form of professional development that is most favored by teachers in TALIS participating countries and reported that 92.6 % of teachers have participated in it. Next, the participation rate for activities such as lectures, workshops, and reading technical books was high. Korean teachers responded that they also participated in informal conversation, lectures, and workshops the most in order to enhance their professional development. However, 77.7 % of teachers in TALIS participating countries responded that they read technical books, whereas only 52.5 % of Korean teachers reported the same. Korean teachers had a higher rate of participation in individual and group research, mentoring and observing coworkers, and visiting other schools for observation than the teachers in TALIS participating countries. Given this result, it seems in-service teachers in Korea have determined that practice-centered training programs are more effective than theory-centered training programs in bringing about changes in teaching.

5.4 Conclusion

Competency standards for teachers have changed in accordance with the times, and accordingly, the contents of teacher education have changed as well. In today's information age, the status and role of teachers are dramatically changing as the learning situation enlarges and the learners' initiative strengthens. In addition, as the movement of school education reform in accordance with social change in the twenty-first century demands changes in teachers, teacher training programs should also change accordingly. Matos et al. (2009) largely divided in-service teacher training programs into training models and participation models. Here, a training model is "a model focused primarily on expanding an individual repertoire of well-defined and skillful classroom practice" and is assimilated to the acquisition metaphor for learning by Sfard (1998). The participation model enhances teachers' professional development by requiring them to participate in professional development activities run by universities or educational institutions and activities in which they can observe their own teaching and discuss learners' responses. It is assimilated to the participation metaphor for learning by Sfard (1998).

Teacher training programs in Korea are advancing toward a participation model as their focus evolves from content knowledge to pedagogical content knowledge; that is, they are changing from lecture-centered training programs with the goal of improving teaching to various forms such as workshops, teacher community, and cyber distance training programs. In 2013, the Ministry of Education presented the goals of teacher training programs, which are, first, a training curriculum for competence in teacher education policy; second, the substantiality of training programs for professional development; third, an increase in the efficiency and utilization of teacher training programs; and fourth, diversification of the training methods, thereby emphasizing pedagogical content knowledge for educational contents to ensure the substantiality of training programs. As the Ministry of Education mentioned workshops, case presentations, group discussions, and site visits for the diversification of the training methods, it presented goals equivalent to the direction of the recent changes in teacher training programs. Further, as explored in Sect. 5.3, there was the attempt to improve teaching through the activities of voluntary associations of Korean teachers, and the government has accepted this as policy, thereby accelerating the direction of teacher training programs.

OECD (2005) reported that teachers in Korea have a low rate of participation in training programs. In response to this announcement, in 2005, the government changed the policy, under which the training programs had been voluntary, to make them mandatory. As a result, the number of distance educational institutes has dramatically increased. According to OECD (2009), about 93 % of teachers in Korea participated in training programs, a number higher than TALIS's average of 87 %, and the average number of training participation days was about 33 days, higher than TALIS's average of about 18 days. Furthermore the Ministry of Education and Science Technology (2010) enacted the development of teaching competence into a law. A teacher evaluation system by students and parents was adopted. Teachers whose student evaluation is 2.5 points or less out of a total of five points must undertake 30 h of training. Clearly, the government's policy helped to increase teachers' participation in training programs. Moreover, in 2011, it was stipulated that teachers should complete over 60 h of on the job training a year, to be reflected in their performance-based pay. Accordingly, it is expected that more teachers will participate in training programs in the future.

However, the increase in distance training programs may not be very helpful in enhancing teachers' professional development. As analyzed in Sect. 5.3, the training courses selected by teachers are usually focused on self-improvement, and there are almost no courses on teaching methods for specific subjects. Because the programs offer one-time training, the formal increase in training will not bring about changes in the current teaching methods. For teachers' professional development to effectively increase the quality of education, the most important elements are active participation by teachers who directly provide instruction and independent willingness. Particularly, as relatively excellent students become teachers in Korea, they earn degrees to develop their own capabilities or form groups specialized by subject. The government must accept this individual-centered training and provide support for teacher communities at the government level to achieve continuous collaborative professional development. If content knowledge and pedagogical content knowledge are balanced and the training method of blended learning is actively used, teacher training programs would be helpful to increase teachers' professionalism.

References

Ball, D. L., & Cohen, D. K. (1999). Developing practice, developing practitioners: Toward a practice-based theory of professional education. In G. Sykes & L. Darling-Hammond (Eds.), *Teaching as the learning profession: Handbook of policy and practice* (pp. 3–32). San Francisco: Jossey Bass.

Cho, K. (2004). Focusing on distance education: A study on improving in-service teacher education in social studies education. *Theory and Research in Citizenship Education, 36*(1), 209–230.

Cobb, P., & McClain, K. (2001). An approach for supporting teachers' learning in social context. In F. L. Lin & T. Cooney (Eds.), *Making sense of mathematics teacher education* (pp. 207–232). Dordrecht: Kluwer.

Cobb, P., McClain, K., Lamberg, T., & Dean, C. (2003). Situating teachers' instructional practices in the institutional setting of the school and school district. *Educational Researcher, 32*(6), 13–24.

Cochran-Smith, M., & Lytle, S. (1999). Relationships of knowledge and practice: Teacher learning in communities. *Review of Research in Education, 24*, 249–305.

Doyle, J. (1991). Rational belief revision (preliminary report). In E. Fikes & E. Sandewall (Eds.), *Proceedings of the Second Conference on Principles of Knowledge Representation and Reasoning* (pp. 163–174). San Mateo, CA: Morgan Kaufmann.

Education Reform Commission. (1996). *Report on new education reform*. Education Reform Commission.

Elbaz, F. (1981). The teacher's practical knowledge: Report of a case study. *Curriculum Inquiry, 11*(4), 43–71.

Elbaz, F. (1983). *Teacher thinking: A study of practical knowledge*. New York: Nichols.

Even, R., & Ball, D. L. (2010). *The professional education and development of teachers*. New York: Springer.

Greenberg, J., & Baron, A. (2000). *Behaviour in organizations*. Prentice Hall.

Hill, H. C., Rowan, B., & Ball, D. L. (2005). Effects of teachers' mathematical knowledge for teaching on student achievement. *American Educational Research Journal, 42*(2), 371–406.

Jeong, J., Kim, H., & Kim, J. (2009). Analysis of the current status of curriculum in distance educational training institute. *Journal of The Korean Association of information Education, 13*(2), 135–144.

Jung, C., Kim, W., Park, B., Lew, H., Shin, H., Lee, J., et al. (1994). A study on the improvement of in-service education program for mathematics teachers. *The Journal of Educational Research in Mathematics, 4*(1), 71–97.

Kazemi, E., & Franke, M. L. (2004). Teacher learning in mathematics: Using student work to promote collective inquiry. *Journal of Mathematics Teacher Education, 7*, 203–235.

Kim, H., & Kim, E. (2013). *An analysis of distance education center in terms of the professional development-focusing on the theme of the program*. Korea Education & Research Information Service: RM 2013-2.

Kim, Y., & Kim, H. (2005). An ethnographic study on the in-service education for the first-class teacher certificates. *Anthropology of Education, 8*(2), 1–30.

Korea Education and Research Information Service. (2013). *2013 Research and Analysis on Consumer Satisfaction of Distance Education Institute*. Research Report 2013-6.

Korean Educational Development Institute. (2013). *Korea and World Education through Statistical Resources*. Statistical Resources SM 2013-02-02.

Kwon, O., & Ju, M. (2012). Standards for professionalization of mathematics teachers: Policy, curricula, and national teacher employment test in Korea. *ZDM: The International Journal on Mathematics Education, 44*, 211–222.

Kwon, O., Park, J., Park, J., & Cho, H. (2014). Designing and implementing professional development program of multi-tiered teacher community: Joint collaboration between teachers and PD program developers. *The Mathematics Education, 53*(2), 201–217.

Lee, H., & Jang, M. (2012). On the analysis and policy alternatives of an in-service teacher training on mathematics education. *Journal of the Korean School Mathematics Society, 15*(1), 171–182.

Lee, Y., Yu, H., & Choi, S. (1993). *A study on the improvement of inservice education of teacher (INSET) in Korea*. Korean Educational Development Institute RR 93-09.

Little, J. W. (1993). Teachers' professional development in a climate of educational reform. *Educational Evaluation and Policy Analysis, 15*, 129–151.

Matos, J. F., Powell, A., Sztajn, P., Ejersbo, L., & Hovermill, J. (2009). Mathematics teachers' professional development: Processes of learning in and from practice. In R. Even & D. L. Ball (Eds.), *The professional education and development of teachers* (pp. 167–183). Springer.

Ministry of Education and Science Technology. (2009). Teacher training certification for the evaluation of training institutes. Retrieved October 16, 2014, from http://www.moe.go.kr

Ministry of Education and Science Technology. (2010). *2010 Development of teaching competence standard manual*. Seoul: Ministry of Education and Science Technology.

Ministry of Education. (2013). *2014 Direction for implementing the teacher training*. Retrieved October 16, 2014, from http://www.moe.go.kr

OECD. (2005). *Teachers matter*. Paris: OECD.

OECD. (2009). *Creating effective teaching and learning environments: First result from TALIS*. Paris: OECD.

Park, S., & Moon, G. (2009). *A study on improvement of implementation of 2007 revised mathematics curriculum for school education competency*. Korea Institute for Curriculum and Evaluation RRC 2009-4-1.

Park, K., Jeong, Y., Kim, H., Kim, D., Choi, S., & Choi, J. (2010). *Research project on the Mathematics Education Development Plan for primary and secondary school in Korea, 2010*. Korea Foundation for the Advancement of Science and Creativity: Policy Research 2010-20.

Sami, F. (2013). South Korea: A success story in mathematics education. *MathAMATYC Educator, 4*(2), 22–28.

Sfard, A. (1998). On two metaphors for learning and the dangers of choosing just one. *Educational Researcher, 27*(2), 4–13.

Shin, H., & Jeon, S. (2008). The analysis of teacher professional development system from the perspective of historical new institutionalism. *The Journal of Research in Education, 32*, 167–192.

Shulman, L. S., & Shulman, J. H. (2004). How and what teachers learn: A shifting perspective. *Journal of Curriculum Studies, 36*(2), 257–271.

Chapter 6
Profiling Mathematics Teacher Professional Development in Malaysia

Chin Mon Chiew and Chap Sam Lim

Abstract This chapter is aimed to provide an insight of mathematics teacher professional development in Malaysia. In general, there exist two strands on what constitute or perceive as teacher professional development in the Malaysian context. On one strand, the education agencies under the Ministry of Education (MOE) conduct in-service courses and workshops for teachers to cater the requirements and changes in the mathematics curriculum. This is meant to ensure that teachers are competent to teach and deliver what is transpired in the curriculum. On the other strand, research-based projects such as action research and Lesson Study have provided some autonomy and empowerment for teachers to dictate their own professional development. Both strands of teacher professional development would have direct or indirect influence towards teachers' teaching. The discussion includes some contemporary issues pertaining to the mathematics teacher professional development. It was observed that generally, teachers show little interest and commitment towards their professional development even though they are aware of its importance in their teaching. The factors are attributed to the over emphasis of examination, administrators' leadership, teachers' skepticism and lack of structural support in school for teachers' professional development. In recent years, the MOE mandated several policies that aimed to put emphasis on teacher professional development such as making it a compulsory requirement for career advancement. There are also efforts to make professional development a school-based programme through teacher collaboration. These would deem more effective and practical in long term as indicated in the research and literatures that promote and support teachers to be self-committed, motivated, and yearning for professional development.

Keywords Professional development · In-service programmes · Action research · Lesson study

C.M. Chiew (✉)
Institute of Teacher Education, Tuanku Bainun Campus, Penang, Malaysia
e-mail: chiewchinmon@gmail.com

C.S. Lim
Universiti Sains Malaysia, Penang, Malaysia
e-mail: cslim@usm.my

B. Kaur et al. (eds.), *Professional Development of Mathematics Teachers*, Mathematics Education – An Asian Perspective,
DOI 10.1007/978-981-10-2598-3_6

6.1 Introduction

The Ministry of Education (MOE) has been granted approximately one-fifth of the annual budget by the government. This acknowledged the importance of education in driving and moving forward the nation into a developed status. Based on the past and current development, the MOE has been dynamic in introducing and implementing changes to the education system: policy, curriculum, pedagogy, assessment, and teacher education with the aim to improve and uplift the standard of education in Malaysia. The Malaysia Education Blueprint 2013–2025 is the current reform launched in 2013 to transform the landscape of Malaysian education system.

Nonetheless, despite some major changes in recent years, mathematics achievement was nowhere to the standard desired. The performance of mathematics in the Trends in International Mathematics and Science Study (TIMSS) and Programme for International Student Assessment (PISA) was rather discouraging. In fact, local research studies concurred some consistencies with the results. In a study by the Higher Education Leadership Academy of the Ministry of Higher Education that observed 125 lessons in 41 schools across Malaysia, they disclosed that 12 % of the lessons were classified as high standard, 38 % met satisfactory standard while 50 % of the lessons were observed to be delivered unsatisfactory (Ministry of Education Malaysia 2012).

Similarly, the findings of *Teaching and Learning of KBSM Mathematics Study* by the Federal Inspectorate of Schools (1994) also revealed some weaknesses in mathematics teaching. It evaluated the teachers' teaching as follows: Good—38.8 %, Average—50.5 %, and Weak—10.7 %. Among the weaknesses of the teachers' teaching highlighted are: (i) questioning techniques were not effective to stimulate students' thinking, (ii) students were rarely encouraged to question and voice their opinion, and (iii) problem-solving activities were not systematically and effectively conducted. These reports deduced that the mathematics teachers are yet to be competent in their basic teaching techniques although the knowledge and skills should have been acquired during their teacher education programme. This reflected some weaknesses in the mathematics teaching that need to be addressed. The MOE is aware that standard of teachers' teaching is much related to teachers' quality. As stated in the McKinsey report (2007), "*no education system can exceed the quality of its teachers.*" Therefore, teacher professional development which is associated to teachers' quality is perceived as the panacea to improve the teaching and learning of mathematics.

6.2 The Mathematics Education Context

This part briefly reveals the education system for readers to grasp and understand the mathematics education context in Malaysia. The education system is perhaps the most unique in the world due to its historical and multicultural background. There exist three types of primary school (Year 1 to Year 6) using similar mathematics curriculum but differ in the language of instruction: Malay, Chinese and Tamil language. However, the secondary school (Form 1 to Form 6) uses Malay language as the medium of instruction while some private schools may either use English or Chinese language. In short, Mathematics is delivered in three different languages at primary level but at secondary level, Mathematics is taught in Malay language for the national curriculum. Nonetheless, there was a short stint when Mathematics is delivered in English under the policy of Teaching and Learning of Science and Mathematics in English which started in 2003 but ended abruptly in 2012.

Despite the differences in the language of instruction at primary school level, it is vital to note that mathematics teacher education programme for both the primary and secondary school is solely in Malay language except for a brief period under the policy of Teaching and Learning of Science and Mathematics in English. As such, some practicing mathematics teachers, in particular the Chinese and Tamil vernacular primary schools are indeed Chinese or Tamil language option teachers but are assigned to teach Mathematics as one of their teaching subject. In other words, they do not acquire any mathematics teaching knowledge and skills during their teacher education programme. In general, the primary school mathematics teachers are graduates from 27 Teacher Education Institutes in the country which confer Bachelor of Education degree. It was upgraded from diploma status since 2007 in line with the MOE's plan to uplift the teaching profession status. Most of the secondary school mathematics teachers are graduates majoring in Mathematics from public universities. In addition, to address the shortage of teachers in the 1990s and early 2000s, non-education graduates were also recruited after attending a one-year education diploma course despite inadequately trained for the teaching tasks.

One of the main issues in mathematics education from the past till present is the discrepancy between teaching practices in the classroom with the mathematics curriculum. Despite several major reforms in mathematics curriculum, it was generally perceived there was little change in actual teachers' teaching practices. For instance, Noor Azlan (1987) disclosed the mismatch of the Modern Mathematics Programme implemented in relation to the intentions of the curriculum developers. He revealed that the activity-based, student-centered, and guided-discovery approach advocated in the secondary school mathematics curriculum was replaced with mainly teacher-centered and "*chalk and talk*" approach in the actual teachings. The notion of effective mathematics teaching was perceived as traditional whole-class teaching strategies with teachers dominating the classroom interaction (Mohd Majid 1997). Lim et al. (2002) acknowledged that *drills and practice* and *memorization* approaches are commonly employed in mathematics teaching. Similarly, Poon (2004) exposed a significant difference between the intention of the curriculum developers and the actual mathematics teaching in classroom. Based on a qualitative case study involving four Form 4 mathematics teachers, she deduced that the constraint was due to the teachers' attitude as they were not aware of the changes and development in the mathematics curriculum. The above scenario definitely posted a serious concern in the context of mathematics teaching as teachers seemed not exhibit the intentions and objectives of the mathematics curriculum.

6.3 Mathematics Teacher Professional Development

Beginning 2013, practicing teachers are required to undergo *Continuing Professional Development* (CPD) course at various stages of their teaching career under the Malaysia Education Blueprint 2013–2015 (Ministry of Education Malaysia 2012). While it was mandated as a requirement for promotion to a higher category or grade, it is aimed to update and enhance teachers the teaching knowledge and skills that are essential to their daily teaching tasks. The first 30-h course of CPD, targeted at novice teachers (below 5 years of teaching service) has four main areas: (i) policy and way forward, (ii) planning and teaching, (iii) managing learning and action research, and (iv) professional values. Other CPD courses for higher grades are being formulated and will be implemented at later stages. It is imperative to note that these CPD courses have reflected the MOE's strategic plan to enhance teachers' professional development. Besides, there was a circular issued in 2005 by the MOE which required teachers to undergo seven days of in-service training in a year. However, when the school administrators were given empowerment, there was quite a variation and flexibility of courses and activities carried out which may not necessary relate to teachers' teaching in the classroom.

To assist and support non-option mathematics teachers who are teaching Mathematics in primary schools, the Teacher Education Division (TED) conducts 6-week In-service Mathematics Conversion course from 1998 until 2005. The aim was to equip these teachers with basic mathematics knowledge and pedagogy skills. In a case study involving 16 participants in a teacher education institute to investigate the effectiveness of the Mathematics Conversion course, Wong (2003) revealed that the overall course content was indeed relevant to their mathematics teaching. However, the content course of abacus, calculator, Internet, enrichment, and remedial mathematics activities were insufficient due to lack of time and therefore, the participants were not confident to engage it in their classroom teaching.

In general, mathematics teachers would appreciate any courses or programmes that would assist them to teach mathematics effectively. However, the in-service courses though relevant towards their professionalism but offer little help to improve students' achievement would likely be ignored. For instance, the MOE via the Curriculum Development Division and state education departments have conducted workshops and courses on Geometer's Sketchpad (GSP) and graphic calculators since 1990s to mathematics teachers at various levels, yet to date they rarely attempt or use it to vary their teaching practices. Despite numerous GSP workshops conducted, Kasmawati (2006) surveyed that only 2 % of mathematics teachers use GSP in their classroom teaching. Furthermore in 2004, the use of abacus was introduced to the primary school mathematics curriculum and subsequently, workshops were conducted to train teachers with regards to the use of abacus in mathematics teaching. A research study by Abd Rahman (2008) revealed that the usage of abacus by mathematics teachers in classroom was extremely low.

In fact, the MOE has been working tirelessly to equip and enhance mathematics teachers' professional knowledge through various in-service courses. Lourdusamy and Tan (1992) classified these courses into three main types of programmes: enrichment, familiarization, and specialization. They noted that the familiarization programmes made the bulk of these courses which is mainly to inform and update teachers' information pertaining to the changes and requirements in the curriculum or policies in the education system. For instance, mathematics teachers were inducted to the *Primary School New Curriculum* since 1983 and the *Secondary School Integrated Curriculum* in 1989 when the MOE revamped the mathematics curriculum. Almost three decades later, the MOE made another major curriculum reform: *Primary School Standard Curriculum* and *Secondary School Standard Curriculum.* In both these curriculum reforms, mathematics teachers were required to attend in-service courses to being informed and update any changes to the mathematics curriculum.

One particular issue that needs to be highlighted is the approach and implementation of the in-service courses adopted by the MOE. A network of master trainers or key personnel is created at national and state levels to coordinate the in-service programmes. Sometimes they designed the modules of training to ensure that the courses will be systematically conducted. These trainers or facilitators are usually lecturers or excellent teachers identified and specially trained by the MOE. In turn, they conduct the courses at state and district levels. In practice, a teacher from each school will be required to attend the course and on his/her return to the school, the teacher will conduct the same content to his/her peers. This approach is known as *cascade strategy* or *multiplier effect* and has been widely adopted due to logistic, time, and financial constraint. One major criticism or weakness of this approach is the information dilution. For example, a two or three-day workshop or course at higher level may be reduced to merely two or three-hour by the teacher at school level. Obviously, the information or impacts received at the lower end are greatly reduced and diluted.

In 1997, the Smart School Project was launched by the MOE to put more emphasis on the application of *Information Communication and Technology* (ICT) in pedagogical practices. The In-service Smart School Course which commenced in 1999 was conducted at 16 teacher education institutes. Mathematics teachers were equipped with a wide range of knowledge and skills on computer hardware, software, networking, multimedia, Internet, and integration of ICT in teaching and learning. This 14-week In-service Smart School Course was funded by the World Bank. A tracer study using questionnaire involved 427 respondents to study if teachers could apply the knowledge and skills acquired during the course to their work in schools. The findings of the study showed that teachers are able to apply the acquired knowledge and skills to prepare and implement in their teaching and learning. On the whole, teachers have positive attitude towards the use of the acquired knowledge. This study concluded that the majority teachers are able to apply the knowledge and skills in planning and implementing teaching and learning in the classroom (Teacher Education Division 2003).

In a related case study involving 46 participants in one of the teacher education institute, Badrul Hisham (2001) admitted that although teachers' skills in ICT and their readiness to improve their ICT skills were enhanced, the study was unsure how it will impact on teachers' teaching. In another study conducted by Supramaniam (2002) involving 55 participants from 29 schools who had attended the 14-week In-service Smart School Course, it revealed that 58.2 % of the participants did not conduct the course to the school teacher upon their return from the course as required. Although the majority participants agreed that the Smart School Curriculum is positive to promote students' learning more effectively, they claimed that the course curriculum was heavy and hence, they are not willing to train other teachers in their schools. The duration of the 14-week Smart School Course was later reduced to 8 weeks, subsequently to 4 weeks and finally, the course discontinued in 2010.

The Professional Development Course for Malaysian Master Trainers in Mathematics [Master Trainer Development Programme (MTDP)] was a 4-week course. It was jointly organized by the Teacher Education Division and SEAMEO-RECSAM, Penang. The main aim of the course was to provide the master trainers in mathematics the opportunities to be exposed to the various innovative and effective teaching and learning approaches in mathematics and on the innovative use of Information and Communication Technology (ICT) in the teaching and learning of mathematics. The course content was rather comprehensive: (i) issues and trends in mathematics education for the twenty-first century, (ii) effective and innovative teaching and learning approaches in mathematics, (iii) using ICT in teaching and learning of mathematics, (iv) instruction and instructional materials for innovative and effective teaching and learning in mathematics, and (v) assessment for innovative and effective teaching and learning in mathematics. This course was designed for excellent mathematics teachers and Mathematics Heads of Panel of schools. The MTDP programme was however short lived, started in 2002 and ended in 2004.

As mentioned earlier, the Mathematics curriculum made a drastic transform when the medium of instruction was changed to English in 2003. The *English for Teaching Mathematics and Science* (ETeMS) course was embarked nationwide to support and assist the teachers to be competent and proficient in the language. A hefty sum was set aside to develop mathematics teachers' linguistic skills, curriculum materials, laptops, LCDs as well as an incentive of 5 or 10 % to their basic salary every month. However, due to some weaknesses in the implementation, the reverse of the policy was made despite a huge allocation was spent over the years. With regards to the ETeMS course, Noraini et al. (2007) concurred that though the teachers perceived they are professionally prepared to teach mathematics in English, they still need more preparation in overcoming students' difficulties in learning mathematics in English. Although the mathematics lecturers of teacher education institutes showed high level of confidence to conduct in-service courses in English, there is a need for continuous professional development among them (Ramlah 2003).

To further support the teaching of Mathematics in English, the Mathematics and Science Trainer Training (MaSTT) course was designed to raise awareness of the role of Content and Language Integrated Learning (CLIL) in the teaching and learning of Mathematics and Science in English. The MaSTT was initiated by the Hongkong and Shanghai Banking Corporation (HSBC) in collaboration with British Council and Teacher Education Division of the MOE in 2005. In line with the Teacher Education Division mission to cater for the continuous professional development of teachers and lecturers, the MaSTT programme was expanded to reach a bigger pool of Mathematics and Science trainers. However, the MaSTT programme was also short lived and ended in 2009. Of late in 2013, mathematics teachers were inducted to Higher Order Thinking Skills (HOTS) 3-day course when the TIMSS 2013 exposed the students' weaknesses in mathematical problem solving and thinking skills. Yet again, the cascade strategy was employed due to logistic and time constraint.

On a different note, the MOE seems to shift the approach of teacher professional development towards school-based under the Malaysia Education Blueprint 2013–2025 (Ministry of Education Malaysia 2012). Experience and excellent teachers were appointed as *School Improvement Specialist Coaches* (SISC+) based at district education offices. The SISC+ main tasks are to assist and coach teachers focusing on teachers' delivery and students' learning in the classroom. This project or initiative is at preliminary stage of the implementation.

6.4 Research-Based Teacher Professional Development

Besides the conventional in-service programmes through courses and workshops organized by the MOE agencies, research-based projects either by a group of teachers or individual for professional development is another strand. Action research, Lesson Study, Professional Learning Communities or any self-initiated school-based professional development activities constitute this strand.

Although action research seems to incline a self-initiative effort for teacher professional development, it was actually promoted by the MOE in 1993. Funded by the World Bank, the *Programme for Innovations, Excellence and Research* (PIER) was launched to improve four educational areas: innovations in science and mathematics, small and isolated schools, distance education and educational research. Subsequently, the action research programme was reorganized and restructured by the Education Planning and Research Department (EPRD) under the MOE to focus and make effective impact on teachers' teaching and students' learning. To assist, support and promote action research among school teachers, short courses and workshops on action research were regularly conducted. A national level research seminar was held annually since 1993 to encourage teachers to carry out action research and present their findings (Bahagian

Perancangan Dan Penyelidikan Dasar Pendidikan 2008). Since 2001, the EPRD provided some funding to inculcate action research culture among teachers. However, the action research was faced with its own challenges and obstacles. Kim (1997) in a state-of-the-practice review of action research noted several challenges faced by action research teams in the conduct of their projects. She revealed that many small-scale action research projects were often short lived and did not go beyond two to three years. Subahan et al. (2002) conducted a survey to investigate the constraints faced by teachers who attempted to inculcate a research culture in the schools. The study revealed that teachers were lack in their knowledge and skills in action research as well as using it and disseminating the research findings.

To further promote and inculcate action research among teachers, Action Research Course is included as one of the core components in the Bachelor of Education Degree Programme at the 27 teacher education institutes (Chee 2010). However, the main concern of the action research programme is whether it can be sustainably carried out by teachers at the school level. While there has been concerted efforts both in preservice and in-service training to provide teachers with skills and knowledge of doing action research, there remains many challenges to overcome at school level. Motivating teachers remains a key element besides the teachers' workload of administrative tasks that has yet to be resolved. In short, extrinsic motivation may be provided through various incentives by the MOE but it is only through intrinsic motivation and teacher empowerment that the inculcation of the action research culture can be seen.

In recent years, a self-initiative and school-based model of teacher professional development that making its way into the Malaysian scene is Lesson Study. Originated from Japan, Lesson Study as a model of teacher professional development began to gain popularity in the United States since 2000 and thereafter spread to other parts of the world. The growth of Lesson Study should be attributed to research projects led by Prof. Lim Chap Sam of Universiti Sains Malaysia since 2003. Some of the research projects were joint efforts with her colleagues while others with students pursuing their postgraduate studies (such as Goh 2007; Chiew 2009; Ong 2010). These research projects were mainly targeted on mathematics teachers in both primary and secondary schools.

Adopting the model of Japanese Lesson Study, Chiew and Lim (2003) first piloted the Lesson Study with a group of five trainee mathematics teachers who were undergoing teaching practicum in a secondary school. The participants claimed that they gained much confidence and their pedagogical content knowledge was enhanced through Lesson Study process. Encouraged by the positive feedback, as reported in Lim et al. (2005), they initiated a Lesson Study research project in two secondary schools within the same district. The aim was to explore the influence of Lesson Study on mathematics teachers' professional development as well as the feasibility of implementing Lesson Study in the Malaysian context. The findings indicated both positive and negative responses. Among the positive responses were through the group discussions and observing other teachers teach, the participating teachers claimed that they gained and enhanced both their mathematics content knowledge and pedagogical knowledge through their self-reflective practice. In addition, the participating teachers also expressed that Lesson Study has encouraged a collaborative culture that enhances professional collegial bonds within their mathematics colleagues. Nonetheless, the challenges faced by the participating teachers in implementing Lesson Study were (i) time factor, (ii) heavy administrative workload, (iii) reluctance and shyness to be observed by colleagues, and (iv) teachers' attitude and commitment. In another study, Goh et al. (2007) shared their insights about mathematics teachers engaged in Lesson Study process. Goh taught Mathematics in a Chinese primary school while Tan taught Mathematics in a secondary school. Despite two different contexts, their reflection about Lesson Study was "*It is really not an easy matter to run and sustain Lesson Study but it is worthwhile when we are seeing the result and benefits gained*" (p. 578).

Besides these research projects, two doctoral dissertations (see Chiew 2009; Ong 2010) and one Master degree (see Goh 2007) on Lesson Study were completed. Based on his doctoral study on two groups of eight secondary school mathematics teachers each in two different schools, Chiew (2009) revealed the positive influences of Lesson Study process on teachers' content and pedagogical content knowledge as well as reflective practice. Similarly, Ong's (2010) doctoral study involved ten mathematics teachers in two different schools. Her finding was teachers' questioning techniques were enhanced through the Lesson Study process. Goh (2007) also reported positive outcome from eight mathematics teachers in a school. The study revealed that teachers' subject matter knowledge and confidence in teaching mathematics using English as the medium of instruction were enhanced.

The positive reports from the research projects involving Lesson Study were indeed fruitful when the Teacher Education Division (TED) of the MOE showed strong interest in implementing Lesson Study to improve teachers' teaching. In 2011, the TED selected 289 schools nationwide to carry out Lesson Study in four different subjects: English, History, Science and Mathematics. In the following year, the programme was renamed as *Professional Learning Communities* (PLC) and three other strategies were added besides Lesson Study: *Learning Walks*, *Teacher Sharing Session* and *Peer Coaching*. The PLC programme, monitored by the TED had involved another 107 schools in 2012, 300 schools in 2013 and 394 schools in 2014. Based on the positive outcomes despite several challenges, the TED has been actively promoting the PLC as school-based teacher professional development.

6.5 Implications and Conclusion

Our review thus far indicates that in general, the MOE has acknowledged the importance and significance of teacher professional development to improve teachers' teaching. From our observations, despite the limitation of the cascade strategy, the top-down in-service programmes organized and conducted by the MOE would likely to stay due to time and logistic constraints. Our analysis also suggested that most of the professional development programmes and in-service courses were often short lived due to rapid changes in the mathematics curriculum and education system. As a result, teachers developed skepticism towards in-service courses that gradually shaped the perception and mindset of the teachers. In addition, the over emphasis on examination in the Malaysian education system is a major issue that yet to be resolved. Consequently, teachers who attend and engage in professional development activities often perceived it as unrealistic and not practical in actual teaching. Moreover, due to situational context, it is difficult to evaluate the effectiveness of any professional development programmes conducted as the nature is rather long term and developmental.

From the evidence-based studies of teacher professional development that relates to effective teachers' teaching and students' learning, Walter and Briggs (2012) summarized the following seven principles of effective professional development: (i) is concrete and classroom based, (ii) brings in expertise from outside the school, (iii) involves teachers in the choice of areas to develop and activities to undertake, (iv) enables teachers to work collaboratively with peers, (v) provides opportunities for mentoring and coaching, (vi) is sustained over time and (vii) is supported by effective school leadership. These principles outlined seem to be consistent with Lesson Study and Professional Learning Communities.

In our opinion, Lesson Study in many aspects is comparable with literatures of effective professional development. In order to encourage and promote Lesson Study as school-based teacher professional development in the Malaysian context, another issue that may arise is the awareness among teachers seeking professional development. The aspects of being voluntary, lifelong learning and self-initiated by the teachers to improve their teaching strategies remain as challenges that are yet to be realized. Hence, more research efforts are needed to explore how Lesson Study could enhance teachers' professional development that ultimately makes an impact on students' learning of mathematics.

References

Abd Rahman, S. (2008). Penggunaan Abakus Dalam Proses Pengajaran and Pembelajaran di Bilik Darjah. *Jurnal Penyelidikan Pendidikan,* Jld. 1. Jabatan Pelajaran Wilayah Persekutuan Labuan, pp. 59–71.

Badrul Hisham, A. O. (2001). *Peningkatan Kemahiran Teknologi Maklumat di Kalangan Peserta Kursus Dalam Perkhidmatan 14 Minggu Komputer Dalam Pendidikan dan Latihan Guru Sekolah Bestari.* Maktab Perguruan Ipoh Perak.

Bahagian Perancangan Dan Penyelidikan Dasar Pendidikan. (2008). *Buku Manual Kajian Tindakan Edisi Tiga.* Malaysia: Kementerian Pendidikan Malaysia.

Chee, K. M. (2010). *Kajian Tindakan: Dari Proses ke Produk.* Pulau Pinang: UPPA, Universiti Sains Malaysia.

Chiew, C. M., & Lim, C. S. (2003). *Impact of lesson study on mathematics trainee teachers.* Paper presented at the International Conference for Mathematics and Science Education, 14–16 October 2003. University of Malaya, Kuala Lumpur.

Chiew, C. M. (2009). *Implementation of Lesson Study as an innovative professional development model among mathematics teachers.* Unpublished Ph. D. thesis, Universiti Sains Malaysia, Penang.

Federal Inspectorate of Schools. (1994). *Kajian Pengajaran dan Pembelajaran Matematik KBSM, (Jun-Julai, 1992).* Kementerian Pendidikan Malaysia: Bahagian Buku Teks.

Goh, S. C. (2007). *Enhancing Mathematics Teachers' Content Knowledge And Their Confidence In Teaching Mathematics Using English Through Lesson Study Process.* Unpublished M.Ed. thesis, Universiti Sains Malaysia, Penang.

Goh, S. C., Tan, K. A., & Lim, C. S. (2007). Engaging in Lesson Study: Our experience. In C. S. Lim, et al. (Eds.), *Proceedings of the 4th East Asia Regional Conference on Mathematics Education [EARCOME4]* (pp. 574–579), 18–22 June 2007. Penang, Malaysia: Universiti Sains Malaysia.

Kasmawati, C. O. (2006). *Meninjau penggunaan Geometer's Sketchpad di kalangan guru matematik sekolah menengah Pulau Pinang.* Unpublished M.Ed. thesis, Universiti Sains Malaysia, Penang.

Kim, P. L. (1997). The environments of action research in Malaysia. In S. Hollingsworth (Ed.), *Action research: A casebook for educational reform* (pp. 238–243). London: Falmer Press.

Lim, C. S., Fatimah, S., & Tan, S. K. (2002). Cultural influences in teaching and learning of mathematics: Methodological challenges and constraints. In D. Edge & B. H. Yeap (Eds.), *Proceedings of Second East Asia Regional Conference on Mathematics Education and Ninth Southeast Asian Conference on Mathematics Education* (Vol. 1, pp. 138–149). Singapore: National Institute of Education.

Lim, C. S., White, A., & Chiew, C. M. (2005). Promoting mathematics teacher collaboration through lesson study: What can we learn from two countries' experience. In A. Rogerson (Ed.), *Proceedings of the 8th International Conference of The Mathematics Education into the 21st Century Project: "Reform, Revolution and Paradigm Shifts in Mathematics Education", pp. 135–139.*

Lourdusamy, A., & Tan, S. K. (1992). Malaysia. In H. B. Leavitt (Ed.), *Issues and problems in teacher education: An international handbook* (pp. 179–191). New York: Greenwood Press.

McKinsey. (2007). *How the world's best-performing school systems come out on top.* New York: McKinsey & Company. Retrieved December 12, 2015, from http://mckinseyonsociety.com/how-the-worlds-best-performing-schools-come-out-on-top

Ministry of Education Malaysia. (2012). *Preliminary Report Malaysia Education Blueprint 2013–2025.* Ministry of Education Malaysia. Retrieved December 12, 2015, from http://www.moe.gov.my/userfiles/file/PPP/Preliminary-Blueprint-Eng.pdf

Mohd Majid, K. (1997). In search of good practice: A case study of Malaysian effective mathematics teachers classroom teaching. *Journal of Science and Mathematics Education in Southeast Asia, XX*(2), 8–20.

Noor Azlan, A. Z. (1987). *The Malaysian mathematics program: A case study of the difference between design intention and classroom implementation.* Unpublished Ph.D. thesis, University of Wisconsin.

Noraini, I., Loh, S. C., Norjoharuddeen, M. N., Ahmad Zabidi, A. R., & Rahimi, M. S. (2007). The professional preparation of Malaysian teachers in the implementation of teaching and learning of mathematics and science in English. *Eurasia Journal of Mathematics, Science & Technology Education, 3*(2), 101–110.

Ong, E. G. (2010). *Changes in Mathematics Teachers' Questioning Techniques Through Lesson Study Process.* Unpublished Ph.D. thesis, Universiti Sains Malaysia, Penang.

Poon, C.Y. (2004). Kurikulum Yang Dihasratkan Dan Kurikulum Yang Dilaksanakan: Pengajaran Penyelesaian Masalah Matematik KBSM. *Paper presented at Seminar Penyelidikan Pendidikan Kebangsaan Bahagian Pendidikan Guru.*

Ramlah, M. (2003). Keberkesanan Pengajaran Pensyarah Matematik dan Sains dalam Bahasa Inggeris di Maktab-maktab Perguruan: Satu tinjauan. *Jurnal Pendidikan Guru,* Bil. 16. Bahagian Pendidikan Guru, Kementerian Pendidikan Malaysia, pp. 47–51.

Subahan, M. M., Abd. Rashid, J., & Jamil, A. (2002). What motivates teachers to conduct research? *Journal of Science and Mathematics Education in Southeast Asia, 25*(1), 1–24.

Supramaniam, A. (2002). Pelaksanaan Kurikulum Sekolah Bestari dalam Kalangan Guru Lepasan Kursus Dalam Perkhidmatan 14 Minggu (Sekolah Bestari) Di Melaka. *Seminar Penyelidikan Pendidikan 2002*. Maktab Perguruan Perempuan Melayu, 73–128.

Teacher Education Division. (2003). *Tracer study in-service courses under the World Bank Loan (1999–2002)*. Ministry of Education, Malaysia: Teacher Education Division.

Walter, C., & Briggs, J. (2012). *What professional development makes the most difference to teachers?* Oxford University Press. Retrieved December 12, 2015, from http://www.education.ox.ac.uk/wordpress/wp-content/uploads/2010/07/WalterBriggs_2012_TeacherDevelopment_public_v2.pdf

Wong, S. W. (2003). Keberkesanan Pelaksanaan Kursus dalam Perkhidmatan 6 Minggu Matematik Guru Bukan Opsyen Matematik Sekolah Rendah. *Cerana Jurnal Pendidikan, Maktab Perguruan Batu Pahat, Bil., 9*, 6–20.

Chapter 7
Congruence Between Context and Opportunities for Professional Development of Mathematics Teachers in the Philippines

Debbie Verzosa, Maria Theresa Tulao-Fernando and Catherine Vistro-Yu

Abstract Like in many other developing countries, mathematics education in the Philippines is often intertwined with macro problems that arise from the sociopolitical context of schools. We investigate the extent to which preservice and in-service education are able to prepare secondary teachers for teaching mathematics at the level of ordinary classrooms. Our analysis is based on the scholarly literature as well as on in-depth interviews with 22 classroom teachers from 12 of 17 Philippine regions who were accepted in a special credential program. We also discuss the macrostructures that exact considerable influence on classroom teaching.

Keywords Pre-service · In-service · Content knowledge · Mathematical pedagogical content knowledge

7.1 Introduction

Due to teachers' crucial role in improving student outcomes, it is necessary to examine teacher preparation at both the preservice and in-service level. The underlying assumption is that mathematics teachers may not be prepared to structure classroom activities that can facilitate mathematical learning. Indeed, studies have shown that mathematics lessons are dominated by rules and drills (Bernardo

D. Verzosa (✉)
Ateneo de Davao University, Davao City, Philippines
e-mail: dmbverzosa@addu.edu.ph

M.T. Tulao-Fernando · C. Vistro-Yu
Ateneo de Manila University, Quezon City, Philippines
e-mail: mtulao@ateneo.edu

C. Vistro-Yu
e-mail: cvistro-yu@ateneo.edu

B. Kaur et al. (eds.), *Professional Development of Mathematics Teachers*, Mathematics Education – An Asian Perspective,
DOI 10.1007/978-981-10-2598-3_7

and Limjap 2012), and that teachers primarily view themselves as transmitters of knowledge (Golla and de Guzman 1998).

In this chapter, we discuss professional development (PD) from the perspective of secondary school mathematics teachers. If teachers are accountable for poor student performance, we find it necessary to also question whether teachers themselves have adequate mechanisms for professional growth. We investigate whether the conditions under which teachers work constrain their pedagogical choices. Specifically, we explore whether their opportunities for PD are aligned with the competencies that are required to teach effectively in Filipino classrooms.

This chapter consists of three parts. First, we present a brief overview of the preservice education for secondary school teachers in the Philippines. Next, we describe typical in-service (INSET) programs. Integrated within these two sections are the perceived effects of these PD programs in terms of addressing teachers' needs. The third section addresses the macrostructures that shape teachers' practice. Unless otherwise specified, our data was drawn from questionnaires and interviews with 22 secondary teachers from 12 of the 17 Philippine regions. These teachers are recipients of a competitive scholarship program for a master's degree in mathematics education. A copy of this chapter was sent to these teachers to validate their responses.

7.2 Context

The Philippine basic education system is governed and regulated by the Department of Education (DepEd). It is a highly centralized and hierarchical structure. All administrative and educational policies are defined by the Central DepEd office (Bernardo and Garcia 2006). According to its official Web site (http://www.deped.gov.ph), the DepEd is organized into 17 Regional offices, which is further composed of a total of 157 Division offices supervising a total of 40,763 elementary and 7683 secondary schools. Despite efforts toward decentralization and school-based management, the educational system is still very much a top-down bureaucracy serving a large number of widely dispersed schools (Luz 2008). Perhaps it is not surprising that within this model, a cascade model of PD is very common, despite its many shortcomings (Nebres 2006).

Like in many other developing countries, the Philippine educational system is fraught with macro problems, or external conditions that pose permanent challenges to mathematics teachers (Nebres 2006). These include large class sizes (i.e., 80 students), lack of infrastructure (such as classrooms or toilets), inefficient dissemination of educational materials, and a large dropout rate. It is within this context that the low achievement of Filipino students (for example, in TIMSS studies) should be considered.

7.3 Preservice Mathematics Teacher Education in the Philippines

Secondary school mathematics teachers typically undergo a 4-year university course leading to the degree of Bachelor of Secondary Education (BSEd) with a Major in Mathematics. In a report by Ogena and Golla to the 2008 IEA-Teacher Education Development Study in Mathematics (TEDS-M) (Tatto et al. 2008), the BSEd Major in Mathematics is offered by 546 teacher education institutions (TEIs). Its curriculum is guided by the Commission on Higher Education (CHED), which is the government body that oversees tertiary education in the country (CHED 2004; for further details, refer to Vistro-Yu and Villena-Diaz 2009).

As prescribed by the curriculum, preservice teachers specializing in mathematics complete 60 units of content courses. An analysis of the 19 TEIs that participated in the TEDS-M revealed that content courses varied considerably—in fact, there is not one mathematics subject that is taught by all TEIs (Basco et al. 2013). In our interviews, we learned that the content courses typically included college algebra, trigonometry, geometry, statistics, calculus, and to a less extent, more advanced subjects. These courses are comparable with the secondary school topics that the preservice teachers are expected to teach, a reality that has not changed since Golla and de Guzman's study in 1998.

The preservice program also includes 51 units of professional education courses. Typical subjects include foundations of education, measurement and evaluation, guidance and counseling, educational technology, and curriculum development. Notably rare are mathematics pedagogy courses, and in some institutions, are nonexistent. This presents a major gap in preservice teacher preparation, especially because mathematical pedagogical content knowledge has been identified as a weakness among mathematics teachers. For example, in a study involving 61 teachers from three Philippine regions, many teachers depended on mathematical rules and could not produce alternative solutions or explanations (MATHTED 2011).

In the context of a developing country, the notion of pedagogical content knowledge includes the macrostructures that surround the classroom environment (Johnson et al. 2000). As they argue, "the environment in which the teacher works still determines which classrooms strategies are workable and which are not." (p. 186). However, from our data, some teachers felt that preservice training hardly prepared them to confront problems that they eventually encountered in schools, such as large class sizes or poor English language proficiency. As Teacher A opined, educational principles are very idealistic but are not applicable in the field. To cite an example, Teacher B described cooperative learning as a teaching strategy encouraged in his preservice program. This setup, however, is not feasible in a crowded room of 70 students. Teacher C also described the strategy of roaming around the room to provide some guidance. In reality, though, his classroom was so cramped that he could not even walk between desks. Teacher A further lamented that preservice teacher educators are often not exposed to public schools, so educational theories are seldom contextualized in local realities.

Teachers also observed that the content of many education courses are theoretical and have hardly been updated over several decades. Teacher D lamented, "we learned to memorize concepts, people, dates and they did not even teach us how these things were related to the kind of education right now." Teacher A felt that the content is shaped by western perspectives, echoing Vistro-Yu's (2008) position that the mathematics taught in the Philippines can be traced to the mathematics instituted by its colonial masters, with little consideration for the Philippine sociocultural context.

The prevailing disconnect between education courses and the local context is somehow addressed by one or two semesters of practice teaching (practicum). During their practicum, preservice teachers are given tasks such as checking assignments, helping students, acting as a teacher's aide and sometimes handling the classes themselves. Although the classes they handled for their practicum were often not comparable to the classes they would eventually handle as teachers, practice teaching is perceived to be the most useful component of the preservice program, primarily because it allows for a genuine experience of the learning environment.

7.4 In-Service Mathematics Teacher Education in the Philippines

An INSET program is annually incorporated in the school calendar released by the DepEd. This INSET may range from 3 to 5 days, and for some teachers, this is the only INSET program they may experience. However, its main focus is not mathematics because it is typically delivered to all secondary school teachers in one school or in one Division, regardless of the year level or the subject that they teach. The teachers we interviewed reported that they were not consulted regarding the design of INSET programs. Often, they only learned about the INSET structure on the day itself.

This annual INSET is obviously limited in terms of developing mathematics teaching. It focuses on assorted topics, including leadership, speech power, journalism, classroom management, children's rights, HIV, or "*kung ano lang mapagisipan ng* speaker (whatever the speaker thinks of)." At times, the topics are based on re-echoes of national or regional seminars that were attended by a selected group of teachers. Another possible INSET focus is on school administration concerns, such as the management of faculty club funds or computation of grades. The speakers or facilitators are sometimes a teacher selected by the supervisor [usually the *kadikit* (favorite)]. Moreover, a teacher may use the INSET to perform demonstration teaching in return for "points" that will contribute to their promotion.

In such a landscape, mathematically focused INSET is often implemented outside the annual school INSET. For example, Teachers E and F (from two different regions) described their INSET experience where a profiling survey was conducted to assess teachers' subject matter knowledge, and the least mastered topics formed the basis of subsequent INSET programs. Teachers met monthly and discussed mathematical lessons and teaching strategies. Teacher F explained that the program was fully supported by her principal and Division supervisors. There was also a level of professionalism whereby INSET facilitators needed to pass a stringent screening process that included a written exam in mathematics and an interview. Those who passed the screening process underwent a 5-day training seminar on how to be effective facilitators.

Indeed, there is no shortage of large-scale and smaller scale school improvement programs that include in-service development as a major thrust (Bernardo and Garcia 2006). These programs represent efforts toward decentralization of the DepEd, and they offer alternative forms of PD other than the cascade model. Outside of these initiatives, however, INSET programs that focus on mathematics education are mostly short term and sporadic. The teachers we interviewed said that the INSET opportunities often do not follow a coherent long-term objective.

Major curricular changes can also spur national-scale INSET. The teachers' experience of these mass trainings reflects Nebres' (2006) repeated accounts of poor implementation of PD. Training was often cascaded (i.e., "echoed") through several levels before reaching the majority of classroom teachers.

Teachers also reported some disjunctures between their INSET experiences and their classroom context. Some INSET seminars focused on the integration of technology in mathematics teaching, while some presented activities that required the use of an LCD projector. However, it is not sensible to expect that there would be enough of the required equipment in resource-poor schools. Additionally, the classroom culture is not always compatible with the teaching strategies presented during INSET. For example, a major focus of a secondary level INSET program was on critical thinking and exploration, but this was considerably different from the "spoonfeeding" method that their students were exposed to in their elementary school years. As one teacher mentioned, "*ikaw na nga gagawa ng activity, ikaw din ang sasagot* (you designed the activity but you end up answering it as well)." Further, for these strategies to work, teachers are compelled to provide worksheets and other materials for students at their own expense.

Several suggestions were offered as to how an "ideal" INSET program could be designed. The most common plea reiterated by the teachers was for more INSET that relate to the specific content and pedagogy of *mathematics*. For them, INSET programs that focus on mathematics are few and far between. A focus on mathematics content is crucial because the teachers themselves admitted that they have very basic knowledge of some secondary school topics. They also observed that many of their colleagues have misconceptions and are not comfortable teaching topics they had not taught before. They also maintained that some secondary school topics were not included in their preservice training, or that these were not taught with much depth.

Teachers also appealed for more INSET programs that focus on mathematics teaching strategies. They generally wanted to learn strategies for teaching basic concepts such as fractions and signed numbers that impede performance in secondary school mathematics. Many expressed the need to learn techniques to make it "easier" for their students to learn mathematics. Some also discussed the need to learn strategies that can increase their students' motivation to learn. In rural areas, for example, many students do not see any relevance in studying mathematics, especially if they do not have plans to pursue a university degree.

7.5 Prohibitive Macrostructures

Considering the data and the literature, three macrostructures were identified to impede student learning and professional growth. First, poverty is often cited as a major cause of absenteeism or dropping out. In rural areas, half the class may miss school to help their families work in the fields. Parents themselves do not necessarily provide enough support for students to stay in school.

A second prohibitive structure arises from the shortness of professionalism within the educational system. Luz (2008) describes how the educational system is largely credential driven, providing incentives for teachers to pursue graduate degrees relentlessly, even if it means enrolling in graduate schools of dubious quality. Because some INSET programs may focus on topics that are not necessarily connected to teachers' concerns, the main incentive for attendance tends to be the certificate handed out at the end.

The third prohibitive macrostructure is the culture of obeisance and the tolerance for corruption (Bautista et al. 2008). In our interviews, teachers describe how structures prevent them from giving students a failing grade, presumably to improve school performance. If they give a failing grade, they may even be summoned by the principal or Division supervisor. Their teaching abilities are questioned and they are blamed for student failure. While teachers are officially encouraged to help students learn, in reality, they are entrenched in a culture of "mass promotion" and "fake achievement." This opens up a cycle of problems wherein teachers themselves are challenged to teach mathematics to students who have not been adequately prepared to learn the expected competencies. The teachers mention examples of sixth-graders who still cannot read or fourth-year high school students who cannot perform operations on signed numbers. They indicate that one shortcoming of the intended curriculum lies in the assumption that students had mastered elementary mathematics. Insofar as teachers are evaluated on the basis of their students' grades and performance in national assessments, teaching practice will not necessarily reflect a teacher's beliefs of effective mathematics teaching (Vistro-Yu and Villena-Diaz 2009).

7.6 Discussion and Future Directions

Teachers are often implicated for poor student performance. In this chapter, we described teachers' perceptions of how PD opportunities are contextualized in local realities. Likewise, there are continuing, but at times sporadic, efforts to address poor mathematics performance. Thus, much work needs to be done in terms of systematically planning PD programs.

A main area of concern is the apparent lack of mastery of mathematics among future and in-service teachers, and the limited opportunities for developing mathematical pedagogical knowledge. Another pressing need is to curb the extent to which prohibitive macrostructures constrain professional growth. In the context of poverty, professional development is only one of many elements of improving education. Indeed, when an empty stomach is a more obvious learning obstacle, the provision of basic services should be included in any discussion about raising performance.

The flawed reward system and the culture of obeisance undermine the value of education. For as long as salary increases and promotion are based on mere certificates, then PD for teachers will remain a farce. For as long as corruption is tolerated at the administrative level, the teachers will see PD as a directive and not something that they could genuinely desire for their own selves.

Clearly, this has to change. Administrators need to temper expectations and relieve teachers from the pressure of reporting success at all cost. More power needs to be devolved to the Regional or Division offices so that PD programs can be followed up and be better suited to the local context. Additionally, stricter standards such as renewal of teaching licenses can help develop a culture of self-improvement and professionalization.

Perhaps due to the top-down structure of the educational system, teachers view themselves to be passive recipients of mathematics content and teaching strategies. Teachers must recognize that they are co-agents of change. In turn, PD programs must develop teachers' adaptive strategies, so that they can be better prepared to carry out the demands of teaching in challenging situations.

References

Basco, L.V., Dimaculangan, F. A., Gomez, B. G., Nool, N. R. Torino, I., Fortes, E., & Villena-Diaz, R. (2013). *Mathematics curriculum analysis of selected TEIs in the Philippines.* Paper presented at the 9th Biennial MATHTED Conference, Bacolod City, Philippines.

Bautista, C. R. B., Bernardo, A. B. I., & Ocampo, D. (2008). *When reforms don't transform: Reflections on institutional reforms in the Department of Education. HDN discussion paper series (2008/2009)* Retrieved May 7, 2011, from http://hdn.org.ph/wp-content/uploads/2008/10/dp02_bautista_etal.pdf

Bernardo, A. B. I., & Garcia, J. A. S. (2006). School improvement in a centralized educational system: The case of the Philippine basic educational system. In J. C. Lee & M. Williams (Eds.), *School improvement: International perspectives* (pp. 227–244). Haugpauge, NY: Nova Science Publishers.

Bernardo, A. B. I., & Limjap, A. (2012). *Investigating the influence of teachers' pedagogical beliefs and reported practices on student achievement in basic mathematics*. Paper presented at ICME 12, Seoul, Korea.

CHED. (2004). *CHED memorandum order no. 30 s. 2004.* Manila: Author.

Golla, E., & de Guzman, E. S. (1998). Teacher preparation in science and mathematics education: A situation analysis. In E. B. Ogena & F. G. Brawner (Eds.), *Science education in the Philippines: Challenges for development* (pp. 41–77). Quezon City, Philippines: NAST SEI CIDS, UP.

Johnson, S., Hodges, M., & Monk, M. (2000). Teacher development and change in South Africa: A critique of the appropriateness of transfer of northern/western practice. *Compare, 30*, 179–192.

Luz, J. M. (2008). *The challenge of governance in a large bureaucracy (Department of Education). HDN discussion paper series (2008/2009)* Retrieved November 9, 2013, from http://hdn.org.ph/wp-content/uploads/2009/05/dp01_luz2.pdf

MATHTED [Philippine Council of Mathematics Teacher Educators]. (2011). *Training Workshop to Develop Mathematical Pedagogical Knowledge*. Unpublished technical report submitted to Science Education Institute (SEI).

Nebres, B. (2006). Philippine perspective on the ICMI comparative study. In F. K. S. Leung, K.-D. Graf, & F. J. Lopez-Real (Eds.), *Mathematics education in different cultural traditions: A comparative study of East Asia and the West* (pp. 277–284). New York: Springer.

Tatto, M. T., Schwille, J., Senk, S. L., Ingvarson, L., Peck, R., & Rowley, G. (2008). *Teacher Education and Development Study in Mathematics (TEDS-M).* Amsterdam: IEA.

Vistro-Yu, C. P. (2008). Ensuring access to quality mathematics education to overcome inequities: A developing country's perspective. A plenary panel presentation, ICME, Monterrey, Mexico, July 6–13.

Vistro-Yu, C. P., & Villena-Diaz, R. (2009). Teachers' beliefs, instructional practices, and culture: Understanding effective mathematics teaching in the Philippines. In J. Cai, G. Kaiser, B. Perry, & N.-Y. Wong (Eds.), *Effective mathematics teaching from teachers' perspectives: National and cross-national studies* (pp. 183–201). Rotterdam, Netherlands: Sense Publishers.

Chapter 8
Professional Development of Mathematics Teachers in Singapore

Berinderjeet Kaur and Lai Fong Wong

Abstract Since the late 1990s professional development of all teachers, including mathematics teachers, in Singapore is guided and supported by the Ministry of Education and other professional bodies. With the adoption of the Professional Learning Communities (PLCs) framework teachers in schools belong to learning teams. They work and learn collaboratively at the school level through participation in a variety of professional development activities. From the narratives of three mathematics teachers about how they work and learn whilst working collaboratively at the school level it is apparent that mathematics teachers develop themselves through a number of ways, such as using resources like research papers and in-service courses to gain knowledge that helps them in resolving issues they face in teaching and learning mathematics. They may also participate in research projects and lesson study as part of PLCs in their respective schools. Teachers also engage in professional development activities to suit their individual needs. They attend higher degree courses at universities in Singapore and elsewhere. They also participate in professional development activities conducted regularly by the Association of Mathematics Educators, Singapore Mathematical Society and the Academy of Singapore Teachers.

Keywords Mathematics teachers · Professional development · Professional learning communities · Lesson study · Singapore

B. Kaur (✉)
National Institute of Education, Singapore, Singapore
e-mail: berinderjeet.kaur@nie.edu.sg

L.F. Wong
Anderson Secondary School, Singapore, Singapore
e-mail: wong_lai_fong@moe.edu.sg

B. Kaur et al. (eds.), *Professional Development of Mathematics Teachers*, Mathematics Education – An Asian Perspective,
DOI 10.1007/978-981-10-2598-3_8

8.1 Introduction

1997 marks the onset of systematic professional development for all teachers in Singapore schools. At the opening of the Seventh International Conference on Thinking on 2nd June, 1997 in Singapore, the then prime minister in his opening speech noted that:

> We must set up comprehensive mechanisms to continually retrain our workforce, and encourage every individual to engage in learning as a matter of necessity. Even the most well-educated worker will stagnate if he does not keep upgrading his skills and knowledge. Every organisation must first recognise the importance of the matter. It must require that its employees go through regular learning as a routine part of working life (Goh 1997).

Following the launch of the Thinking Schools Learning Nation (TSLN) initiative during the Thinking Conference, the Ministry of Education which is a Ministry of the government embarked on a systematic approach to induct teachers in lifelong learning, placing emphasis on continuous professional development so that schools keep abreast of advances in knowledge and learning both at the national and international fronts.

The second initiative, introduced in 2005, was the Teach Less, Learn More (TLLM) initiative (Ministry of Education 2005). TLLM builds on the groundwork laid in place by the systemic and structural improvements under TSLN, and the mindset changes encouraged in Singapore schools. It continues the TSLN journey to improve the quality of interaction between teachers and learners, so that learners are more engaged in learning and better achieve the desired outcomes of education. TLLM aims to touch the hearts and engage the minds of learners, to prepare them for life. It reaches into the core of education—why we teach, what we teach and how we teach. It is about shifting the focus from "quantity" to "quality" in Singapore's education. It emphasizes "more quality" in terms of classroom interaction, opportunities for expression, the learning of lifelong skills and the building of character through innovative and effective teaching approaches and strategies. It also emphasizes "less quantity" in terms of rote-learning, repetitive tests, and following prescribed answers and set formulae.

Systemic infrastructure has been put in place to support the TSLN and TLLM initiatives. Arising from these initiatives, several specific approaches have also been adopted by teachers to embark on their journeys toward excellence in instructional practices. In the following sections, we describe the systemic infrastructure that is prevailing for teachers in Singapore, how teachers work and learn collaboratively and also how teachers may develop themselves through professional activities and university courses at universities in Singapore and elsewhere.

8.2 Systemic Infrastructure

In support of TSLN vision, as of 1998, all teachers in Singapore are entitled to 100 h of training and core-upgrading courses each year to keep abreast with current knowledge and skills. The Professional Development (PD) is funded by the Ministry of Education. To support teachers in mapping their learning trajectories, in 2005 the MOE implemented an Enhanced Performance Management System (EPMS) (MOE undated). The EPMS is an appraisal system that contains rubrics pertaining to fields of excellence in the education system be it teaching, leadership or senior specialist. These rubrics delineate very clearly the competencies deemed necessary at each level and hence teachers are entrusted with responsibility of their own PD. The entitlement of 100 h of PD and EPMS as an appraisal system for teachers has created a significant buzz amongst them for learning opportunities.

For teachers to work collaboratively at the school level, in September 2005, in support of the TLLM initiative "white space" was introduced (Shanmugaratnam 2005). This is the time-tabled time for teachers during curriculum hours to meet, plan and deliberate on their instructional practices. To provide structure for teachers' collaborative work at the school level, in 2010, the Ministry of Education, unveiled the Professional Learning Communities (PLCs) framework (TDD 2010). This framework encourages the formation of Learning Teams in schools. These teams have the choice of adopting a range of collaborative methods/tools, such as Learning circles, Action research and Lesson study, to improve instructional practice through development in subject content knowledge and pedagogy.

In 2009, the Academy of Singapore Teachers (AST) was formed. The subject chapters at the academy are led by master teachers. The key objectives of the chapter are to (i) raise the professional standard in the learning and teaching of Mathematics, (ii) serve as a focal point for teacher collaboration and networking, and (iii) build a culture of professionalism and pride within the fraternity of Mathematics teachers.

8.3 Professional Development of Mathematics Teachers Through Working and Learning Collaboratively

As a consequence of three main developments, which are (i) the introduction of "white space" in 2005 (Shanmugaratnam 2005), (ii) the Ministry's adoption of Professional Learning Communities (PLCs) model as the preferred choice for collaborative learning in 2009 (Lee et al. 2013), and (iii) the PLCs framework (TDD 2010) that provided a structure for teachers to work collaboratively in 2010 teachers including mathematics teachers have been engaged in professional development by working and learning collaboratively. The framework of the PLCs focusses on three aims—improving student learning; building a culture of teacher collaboration; and addressing four critical aspects of outcomes couched in terms of collective responsibility: What is it that we expect students to learn? How will we know when they have learned? How will we respond when they do not learn? How will we respond when they already know it?

The PLCs framework facilitates the formation of Learning Teams in schools. These teams have the choice of adopting a range of collaborative methods/tools, such as Learning circles, Action research and Lesson study, to improve instructional practice through development in subject content knowledge and pedagogy. In all schools, mathematics teachers belong to learning teams. They work and learn collaboratively at the school level through participation in a variety of professional development activities. We draw on three narratives of mathematics teachers to illustrate how teachers develop themselves professionally through participation in PLCs. The first narrative, shown in Fig. 8.1, is written by the head of mathematics department at school A. From Fig. 8.1, it is apparent that mathematics teachers in School A belong to teams, according to the grade levels they teach. Their student outcomes guide them in sourcing for areas of concern they would like to address collectively. They appear to work systematically and collaboratively for a period of time, drawing on resources such as readings (research papers, books, on-line materials, etc.) and in-service courses to enlarge their knowledge and pedagogical skills so as to improve the teaching and learning of mathematics through meaningful activities such as purposeful homework and effective questioning.

The second narrative, shown in Fig. 8.2 (Example 2), is written by a lead mathematics teacher in School B. A lead teacher is one who has demonstrated a high level of competence in both mathematical content and pedagogical content knowledge. In addition to their teaching duties they are also responsible for the development of mathematics teachers in their respective schools and other dedicated schools. From Fig. 8.2, it is apparent that a group of mathematics teachers in School B work and learn collaboratively as a PLC through participation in a research project. The project is funded by the Academies Fund of the Ministry of Education and led by professors at the National Institute of Education, Singapore. The lead teacher who wrote the narrative, is also actively involved in the conceptualization and implementation of the project. It is evident from the narrative that such projects reflect a gradual shift in the centre of gravity away from the University-based, "supply-side", "off-line" forms of knowledge production conducted by university scholars for teachers towards an emergent school-based, demand-side, online, in situ forms of knowledge production conducted by teachers with support from university scholars. Teachers participate, in the project, as a team of four or more from a school. This allows teachers to work collaboratively at the school level to integrate their new knowledge acquired whilst participation in the project into their classroom practice. It is understandable why teachers participate in the project as its goal is in line with improving student teaching and learning which is at the heart of PLCs. Whilst the project facilitates the acquisition of new knowledge and integration of the knowledge into classroom practice, it also facilitates their participation in two PLCs, one at their respective school level and another at the project level.

As part of the PLCs, in my school [School A] mathematics teachers belong to teams according to the grade levels they are teaching, for example grades 7-8 or grades 9-10. I belong to the grades 7-8 team and there are five of us in the team. Each year we examine the student outcomes at the end of the academic year and list two areas that we would like to focus on in the coming year. For example in Nov 2014, the two areas we identified were: effective questioning and meaningful homework. We planned to work on each of the two areas for half of the year 2015.

For the first half of the year we focused on meaningful homework. We drafted our plan for the weekly meetings. First we sourced for readings. All of us read the readings that we could source for homework. After reading and discussing the readings for three weeks, we tried to clarify "what is meaningful homework". After much discussion, we defined it to be homework that would:

- engage our students in consolidating their learning during mathematics lessons (developing skills, revisiting properties, using the concepts and reinforcing the representations),
- experience national examination types of questions, and
- solve non-routine mathematical tasks.

We also tried to work out the proportion of each type of tasks that were appropriate for one homework assignment and the duration of each homework assignment. Finally, we took our homework assignments that we were giving our students and examined them in light of what we had discussed and defined. The exercise led to a significant awareness of how we blindly assigned homework, which at time was only to hone procedural fluency of skills. The team gained much from their shared investigation and certainly worked and learned collaboratively.

In the second half of the year we took a slightly different approach and all five of us enrolled for an in-service course conducted by Prof Berinderjeet Kaur on Effective questioning and facilitation techniques for secondary mathematics teachers at the National Institute of Education, Singapore. After attending the 4 sessions of 3 hours each during out of school hours, we resumed our discussion of how we could integrate our learning into our classroom practice during our PLC meeting times. During these meetings we took turns to plan a hypothetical lesson and focused on what questions we would ask as the lesson unfolded, so that the questions were effective in "getting students to articulate their thought processes". Once again we worked and learned collaboratively. The exercise led us to be aware of how we can move students' thinking up the rungs of the cognitive ladder, from merely asking them to recall to justifying their responses.

Fig. 8.1 Example 1—Activities of a PLC of mathematics teachers in School A

Since the setting up of the Centre for Research in Pedagogy and Practice (CRPP), in 2002, at the National Institute of Education, Singapore projects similar to the one described in example 2 have had a significant impact on the professional development of teachers. Two such past projects are the Enhancing the Pedagogy of Mathematics Teachers (Teaching for Reasoning and Communication) (EPMT-RC) project (Kaur 2009, 2011) and the Think-Things-Through (T^3) project (Yeap and Ho 2009). The aims of the EPMT-RC were three fold: to equip teachers with knowledge about mathematics lessons that facilitate reasoning and communication, support teachers in integrating their new knowledge into classroom practice and contribute towards the development of fellow teachers. The project involved both primary schools and secondary schools mathematics teachers. The deliverables of this project, namely resources crafted by teachers (Kaur and Yeap 2009a, b; Yeap

In my school [School B], the mathematics teachers work and learn collaboratively through Professional Learning Communities (PLCs), level meetings and participation in research projects with NIE professors. One current research project is the Enhanced Pedagogy of Mathematics Teachers (Teaching for Metacognition) with Professor Berinderjeet Kaur and Toh Tin Lam from the National Institute in Singapore.

We work and learn collaboratively in the current project we are participating in. This year, 2015, the mathematics teachers in my school, together with teachers from six other schools in the N6 cluster, came together every Monday, for 6 weeks, to focus on how metacognitive strategies can be infused in the teaching and learning of mathematics, and how a metacognitive culture can be developed in our mathematics classrooms. Teachers were engaged in professional development as they learnt how to craft knowledge building tasks from typical performative tasks found in textbooks and to infuse 10 suggested metacognitive strategies in their lessons, so as to facilitate explicit articulation of students' thinking and encourage students' construction of mathematical knowledge both independently and collaboratively. The tasks crafted were shared with everyone for critique and further refinement.

Teachers were also introduced the concept of Teacher Noticing and were encouraged to video-record their lessons that they later, as a team, reviewed and reflected upon through four lenses – teaching, learning, task and participation. One of my lessons was recorded and viewed by the teachers who in turn gave constructive feedback on how I could further improve my lesson. In addition, teachers learnt a novel tool – lesson play, to write a lesson or part of a lesson in the form of a script for dialogue, featuring imagined interactions between the teacher and students. In the process of writing their lesson plays, teachers' pedagogical and subject matter knowledge were enhanced as they came together to discuss on the mathematical and pedagogical dimensions of teaching, focusing on aspects of practice such as teaching moves and classroom discourse. The teachers' thinking was made visible in the process. Besides crafting knowledge-building tasks and adopting metacognitive strategies to engage students in their learning, teachers also learnt to shape their classroom discourse and interactions by establishing mathematical and socio-mathematical norms that teachers discussed and collectively agreed upon.

Finally the teachers in each of the seven schools worked together to plan and enact lessons designed using the knowledge they had acquired thus far, with the goal of developing metacognitive skills amongst our students. These lessons were video-recorded and showcased in a meeting of all seven schools, presenting segments of the lessons that showed teachers developing metacognition using one or more of the metacognitive strategies identified. Again, peer feedback was sought from the audience who viewed through the four lenses of teaching, learning, task and participation. With the consolidated peer feedback, teachers in each school got together, with guidance from the researchers, reviewed the enactment of their planned lesson. During the review, they wrote a narrative of the lesson highlighting segments of lessons where metacognitive strategies that were evident and also missed opportunities. Teachers were also encouraged to write a journal about their learning of the process of planning, enactment and review. Teachers' learning was the richest at this stage as they learnt that a good lesson comprised careful planning as well as enactment. After addressing the gaps in the enacted lesson, teachers once again worked collaboratively to plan for another mathematics lesson for enactment, bearing the lessons they had learnt in mind.

Fig. 8.2 Example 2—PLC of mathematics teachers in School B

and Kaur 2010) have contributed to several school-based professional activities that have had positive impact on classroom practice of many teachers in Singapore.

The T^3 project investigated teacher change when teachers learnt from each other in a professional community, i.e. knowledge–in–practice (Cochran-Smith and Lytle 1999). The project involved primary school mathematics teachers. The teachers were given a set of word problems to use in their lessons. They were also provided with lesson notes to support their use of the word problems such that students

Mini sausages are sold in packs of 6. Buns are sold in bags of 8. We do not want to have either the sausages or buns leftover.	
Part 1 We use a mini sausage with a bun to make a hotdog. What is the fewest number of packs of sausages and buns we should buy?	Part 2 We use 2 mini sausages with a bun to make a hotdog. What is the fewest number of packs of sausages and buns we should buy?
Part 3 We use 3 mini sausages with a bun to make a hotdog. What is the fewest number of packs of sausages and buns we should buy?	Part 4 We use 1, 2 or 3 mini sausages with a bun to make a hotdog. What is the fewest number of packs of sausages and buns we should buy?

Fig. 8.3 Word problem from T^3 project (Yeap and Ho 2009, p. 138)

considered the contexts of the problems when solving them. They were encouraged to use the word problems provided by the project in their lessons, and create more word problems for use in their lessons. In their professional learning communities they were also encouraged to discuss with colleagues, about the problems and how to use them in their lessons. No structured training was provided to support their learning. Figure 8.3, shows an example of word problem provided by the project for teachers to work on. The three-year long project engaged teachers in professional development while researching their levels of change, i.e. were the teachers merely ignoring, imitating, integrating or internalizing learning afforded by the project.

There are also projects at cluster levels, initiated by school leaders, which involve several schools with groups of mathematics teachers at the school level working and learning collaboratively. One such project was the Assessment Literacy project of the North 2 cluster comprising 15 schools (Chua 2014). This assessment literacy project involved teachers from the 15 schools coming together as a group to acquire knowledge about good assessment practices for mathematics from an expert, groups of teachers from the respective schools (i.e. the PLCs at the school level) selecting one practice and implementing it across a grade level for an academic school year and examining the feasibility of the chosen practice and its impact on student learning. During the implementation year, teachers worked collaboratively in their respective PLCs in schools. They also met with the expert periodically to seek guidance and clarify their knowledge about the mode of assessment they were investigating. The project culminated with the PLCs at the schools coming together during a conference that was organized by the school leaders to showcase their learning through workshops that they conducted for fellow teachers at the national level.

The third narrative, shown in Fig. 8.4 (Example 3), is written by a mathematics teacher in School C. The teacher has been teaching mathematics for the past 7 years to students. From narrative three, in example 4, it is apparent that lesson study is one of the tools teachers adopt for developing themselves whilst working and learning collaboratively in PLCs. Lesson Study first came to the attention of

In my school [School C], the mathematics teachers work and learn collaboratively through Professional Learning Communities (PLCs). The PLCs are free to decide what ever tool they wish to utilise for their development. I belong to a Lesson study group. One of the teachers (expert) in my PLC has worked very closely with Professor Christine Lee from the National Institute of Education, Singapore. Prof Lee is a pioneer for Lesson Study in Singapore. For the school year, 2015 my PLC was engaged in a Lesson Study. The revised school curriculum of 2012 places heightened emphasis on learning experiences for the construction of mathematical knowledge by learners. The topic we adopted for our lesson study was three dimensional geometry - angles between lines and planes. The expert teacher in our group of 5 guided us as we worked through the lesson study cycle over a period of six months. We invited a master teacher of mathematics from the Academy of Singapore teachers and a professor from the National Institute of Education to be part of the resource panel for the demonstration lesson. The post-lesson critique by teachers and the invited quests contributed towards deepening of our knowledge of spatial visualisation and sequencing of activities to aid student's constructing of spatial concepts.

Fig. 8.4 Example 3—PLC of mathematics teachers in School C

educators in Singapore in 2004, during a conference on Cooperative Learning held in Singapore when researchers from the University of Tokyo and Lewis from Mills college shared with participants how Lesson Study was being used in Japan to develop a collaborative culture amongst teachers engaged in professional development of teachers (Lee and Lim-Ratnam 2014). Adoption of lesson study from Japan by educators in Singapore began around the year 2005 (Fan et al. 2009; Lim et al. 2011). In a research project, believed to be the first on Lesson study in Singapore, conducted by Fan et al. (2009) from 2006–2007, it was found that through the actions of planning, teaching, reflecting, and revising, teacher participants deepened their knowledge and skills which resulted from the diverse community that worked together in the study. It was also found to be a good means of mentoring the beginning teachers by senior teachers in a school. Some examples of school-based Lesson Studies involving mathematics teachers, that have been published, are "Division with remainder: lesson study to promote conceptual understanding" (Fang et al. 2012); "Area of rectangles" and "Equivalent fractions" (Fan et al. 2009).

From the three narratives, we are unable to draw any issues that the teachers face whilst participating in PLCs and also developing themselves. Research by Hairon and Dimmock (2012) about PLCs taking root in schools identified three potential difficulties when PLCs were implemented in schools in Singapore. They were high teacher workloads, ambiguity of PLC processes and their efficacy, and hierarchical system and workplaces.

Lee and Lim-Ratnam (2014) also noted that the

> the implementation of lesson study in Singapore has much support in terms of form and structures, but is lacking in the spirit and substance of *jugyu kenkyuu**. The main difference that we perceive between Singapore lesson study and Japanese lesson study is that in Singapore our focus tends to be on student learning gains rather than making connections with the long-term goals of the school or the national curriculum (p. 58).

**jugyu kenkyuu* is the Japanese word for Lesson Study

8.4 Professional Development Activities to Suit Individual Needs

The EPMS entrusts teachers with the responsibility of developing in their fields of work, specifically teaching in this case. Teachers are guided by their mentors in school and self to pursue professional activities that address their needs. For teachers who wish to pursue further professional qualifications, they may enrol for higher degree courses at universities in Singapore and elsewhere. In Singapore, at the National Institute of Education which is an Institute of the Nanyang Technological University these courses lead to master degrees in Education (Mathematics), Science (Mathematics Educators), Art and Philosophy; and also Doctorates in Philosophy and Education. Others may choose to enrol for relevant short in-service courses, workshops, seminars and institutes. These professional learning activities are conducted by university academics, master teachers and senior teachers.

Professional bodies such as the Association of Mathematics Educators (AME), Singapore Mathematical Society (SMS) and the Academy of Singapore Teachers (AST) are active in providing professional development and learning activities for mathematics teachers on a regular basis. Since 2005, the Association of Mathematics Educators holds an annual conference for mathematics teachers. The conference is thematic and supports the trust of the school mathematics curriculum, as shown in Fig. 8.5, as well as initiatives of the Ministry of Education such as 21st century competencies (Ministry of Education 2010). It is held on the first Thursday of the 4 week long school break in June each year. The date of the conference does not conflict with teachers school work as the first week of the 4 week break is dedicated to the development of teachers. The themes of the past 11 conferences are shown in Table 8.1.

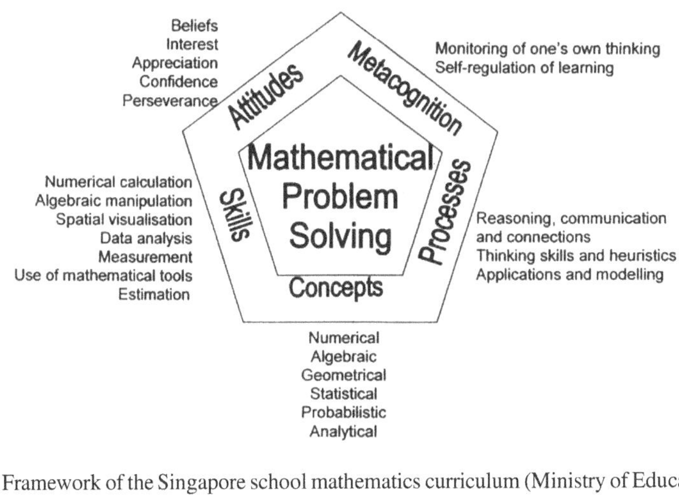

Fig. 8.5 Framework of the Singapore school mathematics curriculum (Ministry of Education 2012)

Table 8.1 Year and theme of AME Mathematics Teacher Conferences in Singapore

Year	Theme of Conference
2005	Assessment
2006	Enhancing mathematical reasoning
2007	Mathematical literacy
2008	Mathematical problem solving
2009	Mathematical applications and modelling
2011	Communication, reasoning and connections
2012	Nurturing reflective learners
2013	Learning experiences in mathematics
2014	Assessment in mathematics
2015	Developing twenty-first century competencies in the mathematics classroom
2016	Empowering mathematics learners

Since 2008, the Association of Mathematics Educators has also published a yearbook that specifically provides for the needs of teachers. Again the yearbook is thematic, as it draws on contributions of scholars to the conferences. It comprises chapters written by renowned mathematics educators from Singapore and the world that are grounded in theory but laden with classroom vignettes and mathematical tasks for teachers to glean knowledge from for their use in classrooms. The titles of the yearbooks published so far are: Mathematical Problem Solving (Kaur et al. 2009), Mathematical Applications and Modelling (Kaur and Dindyal 2010), Assessment in the Mathematics Classroom (Kaur and Wong 2011), Reasoning, Communication and Reasoning in Mathematics (Kaur and Toh 2012), Nurturing Reflective Learners in Mathematics (Kaur 2013), Learning Experiences to Promote Mathematics Learning (Toh et al. 2014), Effective Mathematics Lessons through an Eclectic Singapore Approach (Wong 2015), and Developing 21st Century Competencies in the Mathematics Classroom (Toh and Kaur 2016).

The book Nurturing Reflective Leaners in Mathematics was reviewed by Annie Selden for the Mathematical Association of America (MAA 2013). Selden noted that the subject of the book was metacognition because it is one of five components of the Singapore school mathematics curriculum framework for problem solving and one of the framework's principles for mathematics teaching stated that "teaching should build on students' knowledge, take cognizance of students' interests and experiences; and engage them in active and reflective learning" (Ministry of Education 2012, p. 21). The review highlights that reflective thinking about mathematics does not come naturally to most students, that teachers need well designed tasks that promote reflection, and there needs to be "good teaching and deliberate talk to promote reflection that brings new, higher level perspectives" (Kaur 2013, p. 156). In addition, it states that the book has for school teachers at all levels some ideas of tasks that can be used to induce reflection in students' mathematical thinking. The review by Selden for the Mathematical Association of America affirms the intent of the thematic yearbooks of the Association of Mathematics Education in Singapore.

8.5 Conclusion

It is apparent that since the late 1990s, the Ministry of Education, led by a Minister from the government with a budget that ranks amongst the top three items of the Gross National Product of the country, has guided and provided substantially for the professional development of all teachers in Singapore. Therefore, professional development is an essential component of teachers' lifelong endeavour. Other agencies, professional bodies like the Association of Mathematics Teachers and the Singapore Mathematical Society, too have played a significant role in professionally developing mathematics teachers in Singapore. The sustained support and opportunities for development of teachers, including mathematics teachers, have in the last two decades or so created a culture of lifelong learning at the individual, school and national levels. Therefore, it may be said that no teacher is left behind in developing him or herself. Due to the multitude of opportunities available for mathematics teachers to develop themselves, a lot depends on individuals to set their personal ceiling levels.

References

Chua, P. H. (2014). *N2 cluster mathematics conference on assessment.* Singapore: Unpublished manuscript.

Cochran-Smith, M., & Lytle, S. (1999). Relationships of knowledge and practice: Teacher learning in communities. In A. Iran-Nejad & P. D. Pearson (Eds.), *Review of research in education* (Vol. 24, pp. 249–304). Washington, DC: American Educational Research Association.

Fan, Y., Lee, C. K., & Sharifah Thalha, B. S. H. (2009). Lesson study in mathematics: Three cases from Singapore. In K. Y. Wong, P. Y. Lee, B. Kaur, P. Y. Foong, & S. F. Ng (Eds.), *Mathematics education: The Singapore journey* (pp. 104–129). Singapore: World Scientific.

Fang, Y., Lee, K. E. C., & Yang, Y. (2012). Developing curriculum and pedagogical resources for teacher learning—A lesson study video case of "Division with remainder" from Singapore. *International Journal for Lesson and Learning Studies, 1*(1), 65–84.

Goh, C. T. (1997). Shaping our future: "Thinking Schools" and a "Learning Nation". *Speeches, 21* (3), 12–20 (Singapore: Ministry of Information and the Arts).

Hairon, S., & Dimmock, C. (2012). Singapore schools and professional learning communities: Teacher professional development and school leadership in an Asian hierarchical system. *Educational Review, 64*(4), 405–424.

Kaur, B. (2009). Enhancing the pedagogy of mathematics teachers (EPMT): An innovative professional development project for engaged learning. *The Mathematics Educator, 12*(1), 33–48.

Kaur, B. (2011). Enhancing the pedagogy of mathematics teachers (EPMT) project: A hybrid model of professional development. *ZDM—The International Journal on Mathematics Education, 43*(7), 791–803.

Kaur, B. (Ed.). (2013). *Nurturing reflective learners in mathematics*—Yearbook 2013 Association of Mathematics Educators. World Scientific.

Kaur, B., & Dindyal, J. (Eds.). (2010). *Mathematical applications and modelling*—Yearbook 2010 Association of Mathematics Educators. World Scientific.

Kaur, B. & Toh, T. L. (Eds.). (2012). *Reasoning, communication and reasoning in mathematics*—Yearbook 2012 Association of Mathematics Educators. World Scientific.

Kaur, B., & Wong, K. Y. (Eds.). (2011). *Assessment in the mathematics classroom*—Yearbook 2011 Association of Mathematics Educators. World Scientific.

Kaur, B., & Yeap, B. H. (2009a). *Pathways to reasoning and communication in the primary school mathematics classroom*. Singapore: National Institute of Education.

Kaur, B., & Yeap, B. H. (2009b). *Pathways to reasoning and communication in the secondary school mathematics classroom*. Singapore: National Institute of Education.

Kaur, B., Yeap, B. H. & Kapur, M. (Eds.). (2009). *Mathematical problem solving*—Yearbook 2009 Association of Mathematics Educators. World Scientific.

Lee, D., Hong, H., Tay, W., & Lee, W. O. (2013). Professional learning communities in Singapore schools. *Journal of Co-operative Studies, 46*(2), 53–56.

Lee, K. E. C., & Lim-Ratnam, C. (2014). Exploring the variety and quality in the practice of lesson study in Singapore schools. In K. Wood & S. Sithamparam (Eds.), *Realising learning: Teachers' professional development through lesson and lesson study* (pp. 41–61). London: Routledge.

Lim, C., Lee, C., Saito, E., & Haron, S. S. (2011). Taking stock of lesson study as a platform for teacher development in Singapore. *Asia-Pacific Journal of Teacher Education, 39*(4), 353–365.

MAA (Mathematical Association of America). (2013). *Review of nurturing reflective leaners in mathematics*. Retrieved December 12, 2015 from http://www.maa.org/press/maa-reviews/nurturing-reflective-learners-in-mathematics

Ministry of Education. (2005). *Teach less learn more*. Singapore: Author.

Ministry of Education. (2012). *O-Level, N(A) Level, N(T) level mathematics teaching and learning syllabuses*. Singapore: Author.

Ministry of Education, Singapore. (2010). *MOE to enhance learning of 21st century competencies and strengthen art, music and physical education*. Retrieved September 5, 2015 from www.moe.gov.sg

Ministry of Education. (undated). *Enhanced performance management system*. Singapore: Author.

Shanmugaratnam, T. (2005). *Teach less learn more (TLLM)*. Speech by Mr Tharman Shanmugaratnam, Minister of Education, at the MOE workplan seminar 2005. Retrieved December 12, 2015 from http://www.moe.gov.sg/media/speeches/2005/

TDD (Training and Development Division). (2010). *Schools as professional learning communities*. Singapore: Training and Development Division, Ministry of Education.

Toh, P. C. & Kaur, B. (Eds.). (2016). *Developing 21st century competencies in the mathematics classroom*—Yearbook 2016 Association of Mathematics Educators. World Scientific.

Toh, P. C., Toh, T. L. & Kaur, B. (Eds.). (2014). *Learning experiences to promote mathematics learning*—Yearbook 2014 Association of Mathematics Educators. World Scientific.

Wong, K. Y. (2015). *Effective mathematics lessons through an eclectic Singapore approach*—Yearbook 2015 Association of Mathematics Educators. World Scientific.

Yeap, B. H., & Ho, S. Y. (2009). Teacher change in an informal professional development programme: The 4-I model. In K. Y. Wong, P. Y. Lee, B. Kaur, P. Y. Foong, & S. F. Ng (Eds.). *Mathematics education: The Singapore journey* (pp. 130–149). Singapore: World Scientific.

Yeap, B. H., & Kaur, B. (2010). *Pedagogy for engaged mathematics learning*. Singapore: National Institute of Education.

Chapter 9
Mathematics Teachers Professional Development in Taiwan

Fou-Lai Lin and Yu-Ping Chang

Abstract In this chapter, we present the current status of mathematics teachers' professional development in Taiwan. We first elaborate three policy directions to outline a holistic view on three types of teachers' professional development. The three policy directions are aimed to (1) encourage teachers to take *academic degree program*, especially master degree of teaching, to afford benefits for teaching; (2) provide one-shot or semester-based workshops for teachers' *lifelong learning* for them to catch up on contemporary educational issues and reforms; and (3) incorporate teachers' professional, evaluation, and growth into one system for the convenience of teachers' learning, herein the *network platform*. We summarize the three types with a framework to show the structure of professional development programs involving teachers, their facilitators and contexts. Next, to enhance mathematics teachers' motivation and professions in learning from the workshops, we give two examples of ongoing professional development program, conducted nationally, which assist mathematics teachers in designing tasks and teaching practice, while cultivating their active thinking and learning. Finally, we make a concluding remark on the three types of teachers professional programs in Taiwan.

9.1 Introduction

Teachers Professional Development [TPD] is a complex process but an important avenue to facilitate the quality of teachers, especially in their teaching practice. The process of TPD is commonly understood as teachers learning, teachers learning how to learn, and teachers transforming knowledge and beliefs into teaching practice for the benefit of students' growth (Avalos 2011). How to provide in-service teachers good professional development [PD] has been argued for years

F.-L. Lin · Y.-P. Chang (✉)
National Taiwan Normal University, Taipei, Taiwan
e-mail: yupchang@gmail.com

F.-L. Lin
e-mail: linfl@math.ntnu.edu.tw

B. Kaur et al. (eds.), *Professional Development of Mathematics Teachers*, Mathematics Education – An Asian Perspective,
DOI 10.1007/978-981-10-2598-3_9

Table 9.1 Needs of MTs for PD in academic year 2014

School Levels	Number		
	Teachers (2013)	MTs (2013)	MTs seeking PD in Math (%)
Elementary	88,784	87,306[a]	18,625 (21.33 %)
Secondary (Junior High) (Senior High) (Vocational High)	93,487 (45,604) (33,634) (14,249)	17,248[b]	7335 (42.53 %) (4484) (1902) (949)
Total	182,271	104,554	25,960 (24.83 %)

[a]This number is teachers registered with the speciality in primary education and they can teach mathematics subject
[b]This number is teachers registered with the speciality in mathematics

(Ball and Cohen 1999). Usually, it is viewed that TPD should be flexible and responsive to the changing needs of teachers and professionals (Darling-Hammond and McLaughlin 1995). Since quality mathematics education [QME] differs in countries, and TPD relies on the QME, therefore, the challenges and approaches to TPD differ among countries. In this chapter, we focus on TPD in Taiwan.

In the academic year 2014 in Taiwan, the needs[1] of mathematics teachers [MTs] for PD in the subject of mathematics was around 24.83 % (see Table 9.1) and the number is increasing. To discuss how TPD operates, we intend to introduce the status of TPD with its policy in Taiwan to provide a holistic view of types of TPD programs. Meanwhile in order to present the influential TPD programs in mathematics in Taiwan, we select two ongoing TPD programs as examples to present how local MTs learn to improve their professions.

9.2 Three Types of Teachers Professional Development

There is no doubt that the relationship between policy and instructional practice is close (Cohen and Ball 1990). It means that policy might influence teachers' practice and teachers' teaching might motivate to enact new policy. Regardless of the causal relationship between them both, it needs a bridge to connect them both, and professional development [PD] might be that suitable bridge. In Taiwan though teachers are not specifically regulated to participate in specific TPD in the Education Act, they were required to attend at least 18 h or 1 credit each semester for further education, or accumulated 90 h or 5 credits every 5 years, in the regulation of inservice teachers' further education since 1996. However, this regulation was appealed in 2003. Nowadays, inservice teachers' further education is more

[1]It is supported by the registered numbers, retrieved from the national data (see http://inservice.nknu.edu.tw/Download/103inserviceinvestigations.pdf).
Note: There are around 196,024 school teachers recorded in MOE in 2013.

close to the spirit of TPD which aims to enhance the quality of teachers' teaching practice without any constraints on teachers. However, the situations including personal requirement, school phenomenon/necessity, and educational reforms, push teachers to attend the activities of PD. The activities of PD built by the government can be categorized into three directions: (1) the encouragement and benefit of promotion to motivate teachers to take academic degree; (2) provision of workshops in assisting teachers to understand contemporary educational reform or latest educational issues, e.g., curriculum reform; and (3) construction of a network platform to provide opportunities for teachers to learn conveniently, e.g., the TEACHERNET (https://teachernet.moe.edu.tw) joining various teaching resources, assessments, learning communities, etc. Herein, we discuss three different types of TPD supported by the ideas of those three policy directions.

9.2.1 Type 1: Academic Degree Program

The first direction of TPD encouraging individual teachers pursuing for advanced academic degrees, i.e., master degree or Ph.D. degree, or applying for research program, is under the protection and support of the Teacher Act since 1995. The goal of this direction is for increasing quality education and fostering advanced study among teachers. In latest 20 years, teachers in elementary schools with master degree or PhD degree increase from 1.8 to 43.8 %[2]; in junior high school increase from 4.9 to 40.9 %; in public senior high school increase from 15.4 to 59.1 %; and in public vocational high school increase from 7.8 to 60 %. Though teachers with higher academic degrees increase rapidly in latest two decades, it cannot guarantee the quality education. Therefore, the second direction for improving teachers' expertise is relatively important.

9.2.2 Type 2: Lifelong Learning

In order to provide teachers opportunities to connect contemporary educational issues and reforms to their professions, the MOE authorizes academic institutes, mainly the National Academy for Educational Research [NAER], the Shi-Da Institute for Mathematics Education [SDiME] and universities in Taiwan, to hold workshops regarding to various themes for teachers to learn in different subject domains. The ideas of this direction are close to create *lifelong learning*

[2]Data retrieved from http://stats.moe.gov.tw/files/important/OVERVIEW_M13.XLS are for teachers in elementary school and junior high school, and from http://stats.moe.gov.tw/files/important/OVERVIEW_H11.XLS are for teachers in senior high school and vocational high school.

opportunities for teachers. The workshops can be categorized into two representative groups according to its influences on teachers.

The first categorization is a so-called *learning by listening workshop*. This kind of workshop provides teachers the lectures in general subjects especially related to the latest educational reform issues. However, those lectures might be asked by the government but not necessarily connected to teachers' teaching practice. For example, in latest years, the 12-year compulsory education is one very important and pressing educational reform in Taiwan and is planned to link the connection of all school levels from 2014 to 2020. In order to promote this reform, all the teachers are required by MOE to attend at least five workshops (lectures) in total 18 h, in the themes of *Differentiated Instruction* (tsa-i-hua-chiao-shwei, 差異化教學), *Effective Instruction* (yo-siow-chiao-shwei, 有效教學), and *Multiple Assessment* (duo-yuan-sin-pin-lian, 多元性評量). Those lectures provide issues in general for teachers but rarely related to their teaching practice. Therefore, many teachers make an ironic slogan for those lectures according to the acronym of the first Chinese character of those three themes, *yo-duo-tsa* (有多差) that means *how bad it will be*.

The second categorization is a pioneer in current workshops especially for MTs to participate, named *learning by doing workshop*. This categorization of workshops provides opportunities for MTs to have active and longitudinal, usually semester-based, participation in learning. The aim of these workshops is to transfer the learning style from educators' lectures (learning by listening) to teachers' participation (learning by doing) in designing tasks and teaching practice with the guidance from educators. We later discuss two significant TPD workshops of this categorization in next section as examples.

9.2.3 Type 3: Network Platform

Last, since the information goes rapidly in technological era, it is considered the necessity to integrate learning opportunities for all teachers in Taiwan. Since 2009, the MOE promoted a project to integrate the opportunities of PD for teachers. This project aims to promote a collaborated professional development system TEACHERNET[3] for school teachers. The system provide seven different approaches to assisting in teacher professional growth, including (1) academic courses; (2) resources of digital learning; (3) paradigm; (4) teaching practice and research; (5) professional learning community; (6) professional supporting system; and (7) the system for self-planning professional growth. The former three approaches can be categorized as '*self-regulated learning*', the following three as '*professional collaboration*', and the last one is a synthesis to apply the resources of the former six approaches. Excluding the aforementioned network platform for general subjects supported by the policy, there are also network platform constructed by

[3]See http://teachernet.moe.edu.tw/MAIN/index.aspx.

			Teachers			Facilitator	Context
			Elementary School	Junior High School	Senior (incl. Vocational) High School		
PD Program	Course (Type 1)	Academic Program (incl. degree program)	X	X	X	Educators (incl. Mathematicians, MTEs & MTE-Rs)	Universities
	Workshop (Type 2)	Semester-based workshop	X	X	X	MTEs & MTE-Rs	Schools
		One-shot workshop	X	X	X	Educators (general)	Schools; Academic Institutes
	Research Project (Types 1-3)	Participant	X	X	X	Project Investigators	Schools; Universities
		Independent (anti-network)	X	X	X	Teachers	Mobility

Fig. 9.1 The structure of mathematics TPD in Taiwan

teachers or researchers voluntarily. For example, there is one specific network platform (http://tame.tw/forum.php) for mathematics education. This platform provides all latest mathematics education information in the world and is constructed by a volunteer mathematics educator for decades. This platform is nowadays cooperated with the Taiwan Association for Mathematics Education.

Borko (2004) synthesized the four key elements of making up any professional development system: (1) the professional development *program*; (2) the *teachers*, who are the learners in the system; (3) the *facilitator*, who guides teachers as they construct new knowledge and practices; and (4) the *context* in which the professional development occurs. We structure the mathematics TPD in Taiwan in the following summarized figure (see Fig. 9.1) based on her four elements.

9.3 Examples of Teachers' Active Engagement

TPD Workshops in Taiwan are usually organized by the government as *one-shot* workshops and in a style of *learning by listening*. Not all teachers in school are interested in attending such kind of workshop. Therefore, there are so-called 'professional workshop teachers' assigned by schools to attend such one-shot workshops with very limited efficiency in teaching. How to motivate teachers to participate continually in TPD programs is relatively important and challenging. In this section, we introduce two TPD programs in mathematics, conducted in latest years in Taiwan, which are the style of *learning by doing* in a *semester-based* period, with a lifelong learning spirit in developing teachers' professions by motivating their active participation. Lighten-Up School-Based Program [LUSBP] and Just Do Math (JDM) Project are two prominent and ongoing PD programs for MTs. Both of these two programs are national; however, focus mainly on teachers in the levels of elementary school and junior high school, that is, from grades 1 to 9.

9.3.1 The Lighten-up School-Based Program

The first program is the LUSBP which is a school-based TPD and it provides seven innovative teaching themes in mathematics: *mathematical conjecturing and argumentation*, *diagnostic teaching*, *mathematical literacy and assessment*, *mathematical modeling*, *ICT*, *inquiry-based teaching*, and *reading comprehension*, for schools to choose in each semester since the academic year 2011[4] (see chapter by Lin, Hsu, and Chen in this book for details). The members of the program composed of the MTs, mathematics teacher educators [MTEs], and mathematics teacher educator-researchers [MTE-Rs]. One specific feature of this TPD program is that all the attendants are learners (Lee et al. in review), not only the MTs are learners but also the MTEs and MTE-Rs are learners, though this condition is not emphasized in the program. Here, we present why this program can work successfully in Taiwan by focusing on the attendant MTs' reflections and MTE-Rs' suuport to the MTs with some excerpts as examples. What this program can provide to teachers and how they feel about this program are included.

In attending this program, the attendant MTs gradually understand the essence of the theme they choose and can try to reflect from the connection between their teaching practice and students' learning. They also show their passions in learning and teaching.

> …via these four times workshop, I gradually feel the importance of the process of analyzing and discussing students' minsconceptions and learning difficulties. To solidify the foundation of students' mathematical concepts is far more important than to hurry to finish the curriculum content…the purpose of assessment is to help us to adjust and reflect on our teaching… (one MT from the theme group *diagnostic teaching*)

> (after guiding students explaining their solutions on one geometric problem)…I was impressed with students' solutions and started to believe in their capability that is out of my imagination…I can feel my students' confidence and accomplishment in doing mathematics. I think I am doing the right thing…I also started to appreciate students thinking. (one MT from the the group *diagnostic teaching*)

Moreover, though there are challenges the MTE-Rs meet in the workshop, they try to solve them in the process of TPD workshop. For example, one MTE-R elaborates what problem he met and how he solved it in his theme group:

> After the interaction with teachers and the questionnaire survey, they started to relax a bit from the defensive attitude, and I started to guide them to share…I emphasized it (the workshop) is not instruction, but a process of learning from each other. Theory was less mentioned here (in the workshop). (one MTE-R from the theme of *reading comprehension*)

With more than 3 years experiences of LUSBP TPD workshop for MTs and its extended Light-Up workshop for MTEs and MTE-Rs in Taiwan, it may well conclude that within such TPD workshop all the participants: students, MTs, MTEs,

---

[4]The pilot TPD workshop started in the academic year 2011 before the official program conducted from 2012.

and MTE-Rs, are positioned as learners in several learning communities. This stand supports the policy to motivate teachers' lifelong learning. However, meanwhile it is found that there are tensions faced by MTEs and MTE-Rs, for example, (1) the different identities of in-service MTs, (2) elaborations of theories and principles to MTs, (3) MTs' vulnerability in learning processes, and (4) the weak school support. Moreover, the participant MTs also faces the tensions of designing tasks, tensions on teaching, and tensions on students' learning when they participated in this TPD workshop (Lin 2013).

9.3.2 The Just Do Math Project

There are challenges in mathematics education in Taiwan. Though students out-performed in the large-scale international assessments on mathematics, i.e., PISA and TIMSS, the allocation of high-achieving group and low-achieving group shows a polarization which unveils a serious local mathematics education problem, that is, students' passive learning attitudes and the deficit learning activities of school mathematics (Lin 2015). In order to deal with the challenges of students' passive learning attitudes and the deficit learning activities in school mathematics in Taiwan, the SDiME plans and launches a project to increase students interests and attainments in learning mathematics, called the Just Do Math (JDM) Project. To achieve the aim of this project, it is necessary to cultivate teachers. Therefore, the workshops for teachers are under preparation. The JDM TPD workshop, planned by SDiME and supported by MOE, concentrates on coaching MTEs to design activities and MTs to apply designed activity module to students mathematics learning, in the levels of elementary school and junior high school since 2014.[5]

In this program, the participants are MTs and MTEs. In order to deliver the activity module, MTs have to be certificated as *activity spreader teachers* after a serial coaching workshops. Those activity spreader teachers can choose either Summer/Winter Math Camp or Weekend Math Camp to run those learning activities in schools. Before conduct the project, teachers need deep involvement of designing their specific curriculum and interact actively with the module to make sure the activities can run smoothly. It is expected that there will be at least 2 activity spreader teachers in each school in 1 year. In Taiwan there are around 3500 elementary and junior high schools, therefore, 7000 activity spreader teachers are supposed to be generated in delivering activity module every year. With this expectation that these spreader teachers can coach a new set of spreader teachers in schools, it might ultimately reach the number of 20,000–30,000 in two years.

[5]The JDM program has started its pilot workshop since the academic year 2014. There have been 622 MTs certificated as activity spreader teachers and finished their missions in delivering the activity module to students in this year.

With regard to the roles of MTEs in this program, they have to be instructed by the MTE-R, the director of the SDiME and also the first author of this article, to design the mathematics activities for students who have low attitudes and competencies in learning (see exemplary activities in http://mec.math.ntnu.edu.tw/?page_id=551) and host the workshops for candidate activity spreader teachers to coach them to understand the essence of those activities. Since the JDM aims to equalize students' learning inequality, and provides supports to teachers by the means of cultivating inservice teachers in utilizing designed mathematics activity module to develop students' fundamental mathematical concepts, this program is thought to be a workable way to reform the recent status.

With the pilot experience of the JDM TPD workshop, there are several constructive suggestions in designing learning activities based on the interaction between the MTs, MTEs, and the MTE-R. To better design activity module, the solid ideas kept by the designers are necessary, those are:

- Structualist approach to designing mathematical concepts in embodied tasks
- The connection between iconic/symbolic representation and manupulation
- Starting from students' misconception with a neo diagnostic conjecturing approach
- Familiarity with students' thinking patterns.

9.4 Concluding Remark

The first type of TPD program in Taiwan functions well with the evidence of increasing percentage of teachers having higher academic degrees as the aforementioned trend. However, such demand for the first type TPD is going to be stuck or decreasing in the societal situation. Instead, the needs for enhancing teachers' professions are increasing. Therefore, the second type of TPD program becomes relatively important.

The two exemplary workshops, LUSBP and JDM, provide MTs opportunities to transfer from learning by listening (one-shot workshop) to learning by doing (semester-based workshop). Studying the two TPD workshops, it is found that participant MTs become more active and creative in learning and thinking according to their revised designs in different versions during the workshops and increase their interactions with students in their teaching practices. The communication and interaction between MTs, MTEs, and MTE-Rs in one learning community are becoming more positive as well. However, a fixation on improving students' learning achievement by injecting more tasks still disturbs MTs. Such belief is not easy to ease in participating TPD workshops. Nevertheless, those MTs come to TPD workshop for improving their students' learning and their teaching practice is undoubted. The experiences of semester-based workshops might provide a good model for those one-shot workshops to follow.

Last, though the third type TPD program in Taiwan is in the period of developing, we belief that once the interactions within and between teachers and educators act reciprocally and immediately, the network platforms can work efficiently.

References

Avalos, B. (2011). Teacher professional development in teaching and teacher education over ten years. *Teaching and Teacher Education, 27*, 10–20.

Ball, D. L., & Cohen, D. K. (1999). Developing practice, developing practitioners: Toward a practice-based theory of professional education. In L. Darling-Hammond & G. Sykes (Eds.), *Teaching as the learning profession* (pp. 3–31). San Francisco, CA: Jossey-Bass.

Borko, H. (2004). Professional development and teacher learning: Mapping the terrain. *Educational Researcher, 33*(8), 3–15.

Cohen, D. K., & Ball, D. L. (1990). Relations between policy and practice: A commentary. *Educational Evaluation and Policy Analysis, 12*(3), 331–338.

Darling-Hammond, L., & McLaughlin, M. W. (1995). Policies that support professional development in an era of reform. *Phi Delta Kappan, 76*(8), 597–604.

Lee, Y.-S., Lin, F.-L., Yang, K.-L., & Chen, J.-C. (in review). Exploring MTE-Rs' educating approaches and presented knowledge in design-based workshops for in-service mathematics teachers. *Journal of Mathematics Teacher Education.*

Lin, F.-L. (2013). *An innovative lightening-up program to overcome challenges of Taiwan mathematics education.* Plenary panel at the 6th East Asia Regional Conference on Mathematics Education, March 17–22, 2013, Phuket.

Lin, F.-L. (2015). *Conceptualizing quality mathematics education.* Plenary panel at the 7th East Asia Regional Conference on Mathematics Education, May 11–15, 2015, Cebu.

Ministry of Education. (2014). *'The needs for professional development in specific subject' of school teachers in academic year 2014.* (103學年度中小學教師個人對「在職進修主題細項之需求情形」). Taipei: Ministry of Education. (In Chinese)

Part II
Innovative Professional Development Programs in Asia

Chapter 10
Constraints and Affordances in Bringing About Shifts in Practice Towards Developing Reasoning in Mathematics: A Case Study

Ruchi S. Kumar and K. Subramaniam

Abstract In this chapter, we describe the case study of an in-service teacher, Swati, who participated in a study aimed at supporting teachers to develop resources for teaching mathematics with understanding. We present an analysis of Swati's teaching practice and illustrate how shifts appeared in her practice from procedure based teaching to teaching that supports reasoning. We discuss these shifts in practice in the light of constraints and affordances experienced by the teacher. We claim that the teacher's shift in practice was constrained by tensions experienced with regard to beliefs, tensions experienced in negotiating social norms of mathematics pedagogy and assessment, and limited pedagogical content knowledge. Affordances for exploring reasoning based practices were provided by Swati's reflection on her own practice and beliefs in collaboration with the researcher and peers in workshops, participating in workshop to develop resources for teaching and an alternative image for mathematics teaching and finally, adopting an identity of a teacher focusing on reasoning rather than procedures.

Keywords Teacher practice · Teacher identity · Teacher beliefs · Professional development · Procedure-based teaching · Reasoning-based teaching · Teacher-researcher collaboration

R.S. Kumar (✉) · K. Subramaniam
Homi Bhabha Centre for Science Education, TIFR, Mumbai, India
e-mail: ruchi.kumar31@gmail.com; ruchi@hbcse.tifr.res.in

K. Subramaniam
e-mail: subra@hbcse.tifr.res.in

B. Kaur et al. (eds.), *Professional Development of Mathematics Teachers*, Mathematics Education – An Asian Perspective,
DOI 10.1007/978-981-10-2598-3_10

10.1 Introduction

10.1.1 Practices for Teaching Mathematics in India

Teaching of mathematics in India is largely focused on teaching of procedures for solving textbook based tasks. Clarke (2001) found that Indian teachers expect one right answer, emphasise repetition, listening carefully to examples shown by the teacher and 'explaining' the steps of the procedure again when student gives a wrong answer. The curriculum review focus group on teaching of mathematics raised concerns about the "tyranny of procedure and memorization of formulas in school mathematics" (National Council for Educational Research and Training [NCERT] 2006, p. 6) and the suggestion has been made to focus mathematics teaching on conceptual aspects and processes of mathematics like reasoning and communication. The most common mode adopted for professional development of in-service teachers has been workshops which work on the "cascade model" (Kumar et al. 2015, p. 6) and they have been used to prepare in-service teachers for curricular reform. There has been a parallel effort to reform teacher education by revising the curriculum framework for teacher education to resonate with vision of teaching as portrayed in the school curriculum framework (National Council for Teacher Education [NCTE] 2009). However, there is need to explore alternative models for in-service teacher education which support teachers in developing shared vision of teaching along with providing support for exploration of practices that support understanding.

10.1.2 Models for Supporting Teacher Change in Practice

Several models for supporting in-service teachers' professional development in the context of curriculum reform have been proposed across the world which aim to 'change' teachers' practices. Training model currently used in India assumes that teachers can change their behaviours to adopt new practices proposed by authorities as worthy of replication (Sparks and Loucks-Horsley 1990). Richardson and Anders (1994) discuss how issues of teachers' change are related to power in terms of who drives the change. They argues that teachers resist change enforced by authorities which they fail to make sense of, but continuously undergo voluntary change. Thus she engaged teachers in a collaborative study to help develop identity of autonomous teachers who make informed decisions and chart their own trajectory of change. Clarke (1994) identifies addressing issues of teachers' concerns and soliciting teachers' conscious commitment to participate actively as one of the key elements for professional development. In consonance with these studies, we recognise teachers' agency in driving their own professional development.

Teacher beliefs and knowledge have been considered as important factors governing teachers' practice consciously or unconsciously (Swan 2006). In the context of teacher change, some researchers argue that beliefs act like filters (Pajares 1992) and thus teachers interpret texts and discourse in confirmation of their held beliefs rather than as precursor to reflections. Guskey (2000) on the other hand proposed that teachers change beliefs as a result of success of students in learning. However, researches have also pointed out the relevance of mathematical knowledge for teaching in determining mathematical quality of instruction (Hill et al. 2008). Recognising that teachers' knowledge is situated in classroom experiences and that teachers need to critically reflect on them in order to change practices, Putnam and Borko (2000) discuss how combining multiple contexts like workshop and ongoing support during school year is promising for teacher learning. They argue that workshops can promote "developing different conceptions of mathematics and deeper understanding of mathematical learning and teaching", while teachers' own classrooms can be sites to explore enactment of specific practices (Putnam and Borko 2000, p. 7).

The multiple factors recognised in research literature for influencing classroom practice indicate the need for frameworks for professional development that acknowledge the complexity of the process. Clarke and Hollingsworth (2002) proposed an empirically grounded framework for teacher change which is non-linear and describes change as a process of reflection and enactment in "the personal domain (teacher knowledge, beliefs and attitudes), the domain of practice (professional experimentation), the domain of consequence (salient outcomes), and the external domain (sources of information, stimulus or support)" (p. 950). While this framework explains the process of change that the individual teacher undergoes, we need to also consider that the teacher is situated in a social context. Professional development occurs through the process of peer interaction in a professional development context within schools and outside school settings. Gresalfi and Cobb (2011) described the process of teacher change in terms of development of identities in school and professional development context.

Professional learning communities comprising teachers, teacher educators and researchers, who are engaged in the enterprise of mathematics education, have also been considered promising. They allow bringing varied experience and situated knowledge of students and contexts of teachers into the discourse of the community for reflection and developing insights, while the presence of university educators helps in bringing critical and reflective stance in conversations and bringing research based ideas of teaching and learning into the discourse (Goldenberg and Gallimore 1991; Saunders et al. 1992; Richardson and Anders 1994). Vescio et al. (2008) review of professional learning communities indicates that participation in these communities did lead to change in teaching practice in terms of it becoming more student centered.

10.2 The Study

10.2.1 Theoretical Framework

Drawing on Wenger's theory of 'communities of practice' (1999) and situated learning theory by Matos et al. (2009), this study adopts the view of looking at in-service teachers as professionals who have been engaged in teaching for several years and thus possess situated knowledge of students and the contexts in which they teach. Teachers are thus viewed as knowledgeable members in the community of mathematics educators. We adopt a broader view of 'community' as encompassing teachers educators and researchers along with teachers who all are engaged in the enterprise of analyzing teaching by using experience and evidences from their classrooms and exploring alternatives to support students' learning and improving their practice (Jaworski 2008; Brodie 2012). Even though teachers are members of the community of teachers, they may not have opportunities to communicate and discuss "teaching" in their everyday work and reflect and learn about teaching. Matos et al. (2009) elaborate on how practice based professional development is promising as "the text of teaching serves as the context for teachers to learn about the specific aspects of their labour and reflection is expected to increase teachers' awareness of practice, allowing them to make thoughtful decisions in the immediacy of classroom work" (Sect. 10.1.2, para 2).

Professional development of teachers in a professional learning community encompasses articulating and sharing their beliefs and knowledge with other members of the community and thus participating in the process of knowledge construction by supporting and challenging the views articulated based on their experiences. The professional development is then a social process that of development of professional community comprising teachers, teacher educators and researchers rather than development of an individual teacher who assimilates the materials and lectures transmitted in a professional development program. We view situatedness, challenge and community as three aspects of professional development through which opportunities are provided to teachers to revise, reflect upon and rebuild the knowledge and beliefs held by them and to re-negotiate their identity. In this chapter, we illustrate the process of how one teacher negotiated her identity as a result of participating in a professional learning community in workshops which was extended to the school setting. The teacher's identity was negotiated by critical reflection on beliefs and practice and exhibiting agency in terms of exploring practice for developing reasoning in mathematics as compared to focus on procedures. The research questions addressed in this chapter are:

1. What were the shifts in practice of teacher Swati if any, during the course of the study and across what dimensions did these shifts occur?
2. What were the constraints and affordances for the intended changes in practice which can be gleaned from teacher's reflections and participation in workshop and school settings during the study?

10.2.2 Research Design

The study covered a span of two academic years and consisted of two phases. In the first phase, a ten day long workshop during the summer vacation was followed by collaboration with the teacher in the classroom. The goals of the workshop were strengthening teachers' knowledge relevant to teaching, providing opportunities to articulate and reflect on beliefs and developing a sense of community among teachers, teacher educators and researchers participating in the study. Classroom collaboration was included to identify the take up from the workshop and challenges faced by teachers in implementing intended changes. The second phase included six one-day workshops spread over a period of five months that were held while the teachers were teaching in their schools. The specific topic of integers was chosen to work in workshops and for classroom collaboration with a group of 4 middle school teachers. Swati was part of this group. At the end of the second phase, the teachers' group conducted an extended workshop session on teaching integers for peer teachers from same school system (Fig. 10.1).

We have adopted the case study approach in this chapter as in depth analysis of one teacher will help to identify the different factors that afford or constrain teacher intended change. Extended observations of the classroom are important in building a more nuanced picture of teachers' practice. The particular teacher Swati (Pseudonym) was selected during the first workshop for closer analysis and extended collaboration in school as she represented a 'teacher-in-flux' who was experiencing tensions in her articulated beliefs as a result of curriculum reform. We expected analysis of data from her case to illuminate how professional development context and school setting impact the tensions experienced by her.

Swati was a middle school teacher with a Master's degree in mathematics and a B.Ed. degree. At the age of 42 years, she had a teaching experience of 17 years, out of which the first ten were teaching science and mathematics at primary level. For the last 7 years, she had taught mathematics from Grade VI to X. In her interview she said that she had positive experiences of learning mathematics as a student; she was motivated to be "first in class" in mathematics and getting "excellent" grades from the teacher.

10.2.3 Data Collection

The data was collected in the form of notes made by the researcher of the informal interactions with the teachers and authorities in the school and discussion with Swati after lessons. All the lessons were audio-recorded with the recorder placed on the teachers' table in the front of the class. The researcher also wrote logs in the classroom of the interactions and the personal reflections about the class. A total of 46 classes were observed in 2 phases, details of which are given in Table 10.1 below.

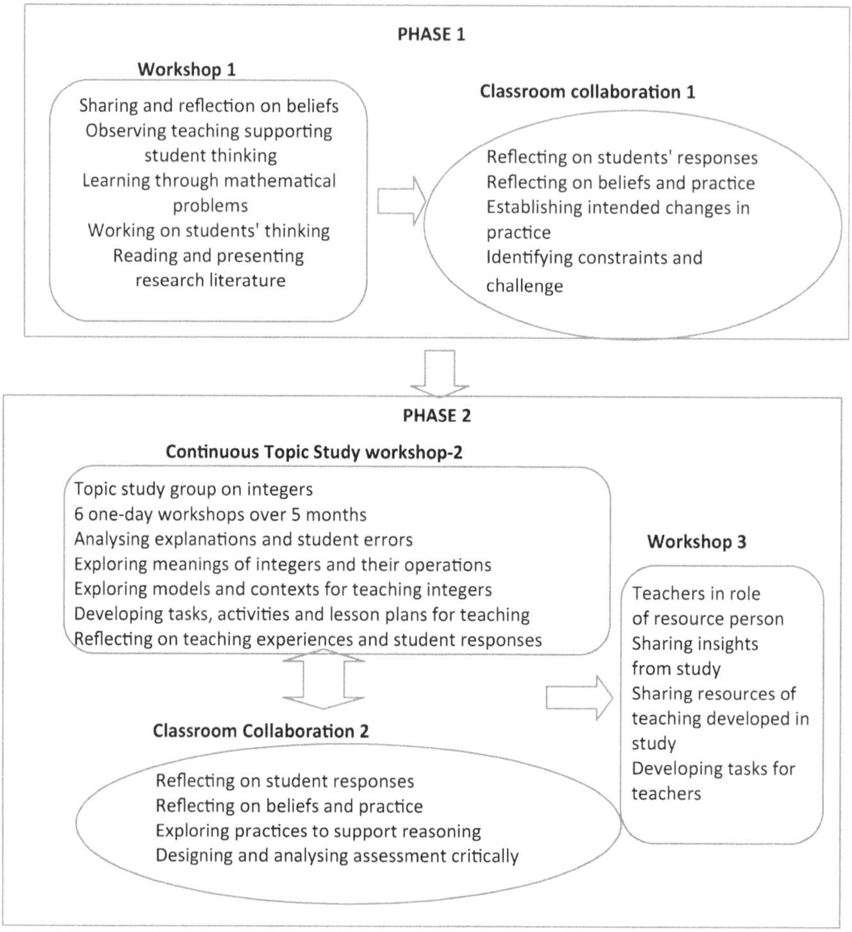

Fig. 10.1 Study design depicting professional development efforts in phase one and phase two of the study

Table 10.1 Classes observed of teacher Swati in the two phases

Phase	Period	Number of lessons observed	Grade	Topics
1	Aug.–Sept. 2009	15	VI, VIII	Algebra, lines and angles, triangle and its properties
2	July 2010	16	VI, VIII	Divisibility tests, quadrilaterals
2	Oct.–Nov. 2010	15	VI	Integers

Table 10.2 List of codes to analyse practices used by teacher in classroom for teaching mathematics

	Aspect of teaching practice	Description	Examples
1	Explanations		
1.1	Procedural explanation	Focus is on the steps of the solution or telling procedure in general to solve a particular type of problem	"First you will have to do, if any you can do multiplication and division and then next you have to do addition and subtraction."
1.2	Conceptual explanation	Focus is on the meanings of mathematical concepts and symbols used	"so plus means increase, minus means decrease… increase is this sign and decrease is this sign."
1.3	Use of representation	Teacher's use of representation other than symbolic like models, contexts or descriptions of situations, visual representations	"When I talk in terms of centigrade it is 10 degrees less than zero. Right? So that is why it is negative and that is why it is less than zero. So all negative numbers are less than positive numbers and even zero."
2	Tasks		
2.1	Textbook based	Teacher gives question from the textbook	"We were doing exercises 3.3. We will take up third question."
2.2	Constructed by teacher	Teacher constructs tasks for students based on her decision as to what is relevant to develop understanding	Teacher asked students to solve $3 + (-2)$ followed by $+3 - (+2)$ on number line and compared solutions.
3	Students' solution of problem		
3.1	Teacher telling steps to follow	Teacher asks student to solve but tells steps of solution to more or less extent	"So here, what we will do first? Multiply… same thing [as earlier problem], we will multiply with any number, any number you tell…"
3.2	Doing on one's own based on examples/steps told	Teacher solves an example problem and shows all the steps first and then asks students to solve a similar problem	"This is the next question. You will not first do 4 + 3… some bracket is also there so you will first multiply with both…"
3.3	Doing on one's own without being told the procedure	Teacher asks students to solve a problem without telling steps	
3.4	Giving explanation and reasons for the solution to others	Teacher asks student to explain answers or give reason for their answers	"Who will explain what she has done here?"

(continued)

Table 10.2 (continued)

	Aspect of teaching practice	Description	Examples
4	Teacher questions		
4.1	Procedure based	Teacher asks the students to give steps of the solution	"What will be the first step?"
4.2	Reasoning based	Teacher asks students to give reasons by asking why questions	"Why? If it is minus then why is it less? You tell."
4.3	Use of representation	Teacher asks question using a representation	"You have to go from 5th floor to 2nd floor. So how will you write mathematical statement?"
5	Teachers' response to students' response		
5.1	Evaluation	Teacher evaluates the answer as right or wrong	"correct, very good"
5.2	Procedure based	Teacher asks for the next step in solution or states the general procedure for the problem	"What we have to minus in this [expression]? Then also, we come back [towards negative direction] and if there is negative integer then also we come back"
5.3	Reasoning based	Teacher explains or asks reason in response to students' answer, does not accept rule based answer and asks for explanation	"Yes, what she said that because –352, it is less than zero. it is 352 less than 0 that is why it is minus… ok it is 352 times less than zero, so zero is more than – all negative numbers are less than zero."
5.4	Giving student autonomy	Asking students to evaluate the answers, asking students to raise their hand for answer being correct without giving her own evaluation	"First everybody said 'yes', he said wrong, so everybody said wrong. Have your own thought."

10.2.4 Data Analysis

Day-wise description of teaching and discussions with the teacher was constructed using the field notes of the researcher and logs of the classroom. The description focused on different practices adopted by the teacher to give explanations, tasks used, questions posed, expectations for solution and teachers' responses to students. Comparison was done between practices adopted by the teacher in the first and the second phases by selecting two lessons from each phase and analysing their transcripts. The turns of teacher talk were reviewed to arrive at emergent codes for denoting teacher's practices, which were later subsumed into categories (Miles and Huberman 1994). For description of codes and example see Table 10.2.

Using this framework of codes, transcripts of two lessons on algebra in the beginning of Phase 1 and two lessons on integers at the end of Phase 2 were analysed, which were taken to represent two ends of the teacher's journey while participating in the study. In the transcripts, each turn of teacher talk was coded using the framework and counts are presented in Table 10.3. For many turns, more than one code was given when, for example, the teachers' response to students' response was procedural explanation resulting in code being given for 'procedural explanation' as well as 'procedural response'.

10.2.5 Results

In the description to follow, we illustrate how Swati's participation in a learning community helped her in exploring new practices in her teaching and negotiating her identity of being a mathematics teacher who emphasises reasoning as a learning goal. In the first subsection, we describe the shifts in practices related to explanations, tasks, questions, students' solutions and teacher's response to students. In the second subsection, we discuss the factors underlying these changes as expressed through Swati's own reflections on moving from being a teacher focusing on telling procedures and rules in class to a teacher exploring and struggling to develop reasoning amongst her students, focusing on the constraints and affordances she experienced in this journey.

10.2.5.1 Shifts in Classroom Practices of Teacher Swati

The shift in Swati's teaching practices can be characterised along the twin continuum of having procedural focus to reasoning focus on one hand and from being teacher centred to becoming more student centred on the other. These shifts have been characterised by analysing practices related to explanations, source of tasks, questions posed by teacher, expectations for students' solutions and how the teacher responded to students which are illustrated in the framework in Table 10.2. In this section, we present analysis of two lessons each from the two phases of the study. The analysis across these five dimensions of practice for these 4 classes is presented in Table 10.3 along with examples from the transcripts. We describe the general observations of the practice in both phases of the study, illustrating with examples from the classes analysed.

Classroom observation of Swati's teaching in the sixth grade in phase 1 indicated that calculations and manipulations of symbols constituted the major portion of teaching with hardly any space for conceptual discussion or use of other representations like concrete materials, models, situations or visual representations. The teacher showed how to solve the problem symbolically on the blackboard, after which she asked students to solve similar problems, sometimes by looking at previously solved questions. The teacher also highlighted the 'points to remember'

Table 10.3 Comparison of practices used by Swati in the first and second phase of the study

		Phase 1		Examples	Phase 2		Examples
		L 1	L 2		L 1	L 2	
1	Explanation						
1.1	Procedural	33	25	"These are only numbers, no variable is there… so these two are like terms. So, you can add only these two. You cannot add this to these. Okay. You can only add or subtract like terms like $4t$ and $5t$."	8	12	"He has done it without using number line… he has used what here- additive inverse. [It is used] when we change the subtraction question into addition question."
1.2	Conceptual	1	0	"When you transpose the term… we have learnt how to do the same thing. [Add] number on both sides, so it is actually +5. So minus $5p$ on both sides, then only $5p$ will cancel…"	2	5	"What he has done, went from zero to 3 and −2. [This] means it is minus so you have to come back… minus also tells you that it has decreased."
1.3	Use of Representation other than symbolic	0	0	None	8	4	"You can always think of a situation through which you can understand.… You have to relate numbers to those situation. Runs, number of runs…"
2	Tasks						
2.1	Textbook based	6	4		1	0	
2.2	Constructed by teacher	0	2		2	6	
3	Student solution						
3.1	Teacher telling steps	19	19	"There is no sign in front of 4. What does it mean? It is plus."	6	0	"See it has become so easy… don't change the sign of first number. Why you don't change this, because you have to subtract this number from this. Then add additive inverse."

(continued)

Table 10.3 (continued)

		Phase 1		Examples	Phase 2		Examples
		L 1	L 2		L 1	L 2	
3.2	Based on examples/steps told	3	3	After solving one example on blackboard teacher said to students, "Do the next one in your notebooks. I will see."	0	1	Student complaining to teacher "Madam, she is solving by looking at that solution."
3.3	Thinking on one's own	2	0	Teacher asking students, "Any other way you can solve [silence] Any doubt? No?"	5	7	Teacher saying to students, "If it is wrong then come and correct it. Why are you looking at him [so-called 'bright' child]? Whatever you think you tell. …"
3.4	Explanation of answer	0	0		0	12	Student's response, "+3 + (−2) it will be −1. So Madam, +3 − (−2) has to be +5… Madam, it has to be different, it cannot be same."
4	Teacher questions						
4.1	Procedural	17	26	"This is 4 + 3t + 6. So, now next what will it be?"	12	10	"How can you convert this [subtraction problem] to addition?"
4.2	Reasoning	9	1	"Whether you have x = 4 or 4 = x, what difference does it make? Is there any difference between the two?"	6	16	After solving 3 + (−2) and (+3) − (+2) teacher asked, "Answer is coming same and numbers are also same but what is the difference between these two?"
4.3	Use of Representation	0	0		2	5	"If you have to go from fifth floor to second floor what button you will press + or − and how many times?"

(continued)

Table 10.3 (continued)

		Phase 1		Examples	Phase 2		Examples
		L 1	L 2		L 1	L 2	
5	Teacher's response to students' response						
5.1	Evaluatory	5	9	"correct, very good"	1	7	"It is perfectly correct. So the answer is −7 [speaking to S1]. You want to change your answer? [to S2]."
5.2	Procedural-explaining or asking steps	23	20	"Which operation should come first?"	4	8	"What we have to minus in this [expression], then also we come back [towards negative direction] and if there is negative integer then also we come back."
5.3	Reasoning based	8	1	After deliberately making a mistake of adding variable and number and student calling it wrong "We cannot add variable and number, why?"	6	20	"I do not want rules. I want you to tell me based on what we have done till now. What is the role of minus operation and minus sign? You have to use both."
5.4	Giving student autonomy	4	0	"Who is saying this is wrong? Raise your hands. Come and do the correct one then."	2	5	"First everybody said yes, he ['bright' child] said wrong so everybody said wrong. Have your own thought."

Legend L 1—Lesson 1, L 2—Lesson 2

like multiplying with both numbers after opening the bracket with the aim of avoiding students' errors. So most explanations were procedural although the teacher did try to give conceptual explanations sometimes like in line 1.2 in Table 10.3. However, the conceptual explanation was not discussed at length. In the conversation after the lesson, the teacher shared that students find the explanation of adding the same number to both sides of the equation more difficult since they have already been taught to use 'transposition' in tuitions or at home.

In Phase 2, the number of instances of procedural explanation is far fewer than in Phase 1 and the teacher focused on eliciting students' methods, explanations and reasons rather than giving explanations. Although she tried giving conceptual explanations, there was a tendency to convert the conceptual explanation into a procedural one. For example, initially, she used the meaning of addition as increase and subtraction as decrease to explain the movement on the number line (a meaning based explanation) but would slip into relating minus sign with going in negative direction (a visual convention based explanation). On the other hand, she did explore the use of contexts, models and activities to discuss meanings of integers and their operations although she believed in the first phase that contexts and concrete models distract students and do not help in learning mathematics. However, here also she would slip into using them procedurally by telling students what to do to get the correct answer, whenever she got wrong answers from students or felt that the students did not understand.

In the first phase, Swati covered the textbook page by page by following the sequence of tasks given in the textbook and doing questions mostly given in the textbook. In fact, her lesson would start by asking students which textbook exercise and question they had done in the previous lesson. Since many students had been going to tuition classes outside schools and doing chapters ahead of when it is done in class, a number of them already knew the procedure to solve and thus the pace of lesson was established by these students. Swati was aware of this in the first phase and cited it as a reason for the lack of student engagement in the lesson. However, in the second phase, Swati consciously made efforts to engage students in understanding by constructing tasks and resources like activities, contexts and models developed collaboratively in workshops with researchers for the teaching of integers.

The questions posed by Swati to students were also procedure based in the first phase while comparatively more reasoning and representation based questions were posed during the second phase. A feature very common in the first phase was teacher asking *narrower* questions (Stein et al. 1996), where the cognitive demand of the questions is so minimal that one is bound to get a correct answer from students. The teacher used a series of such questions to show steps for a solution, e.g. "We transpose 4 in equation to get $3x = 7 - 4$. What is $7 - 4$?". Most questions in the first phase were about asking the next step in the solution or the procedure and the ultimate goal for the series of question was to get to the answer. She did ask some "why" questions but accepted or gave explanation of the question with a procedure. In the second phase, Swati not only asked more reasoning based questions but also posed questions using contexts or models suggesting a shift away from a preference for symbolic representation (see Table 10.3).

With the shift in the type of questions posed, there was also a shift in the expectations that the teacher had of student responses to the problem posed. While in the first phase the students were expected to follow the method told to them by the teacher, in the second phase Swati refrained from telling steps of the solution or showing an example but sought more explanation from students of the answer given by them. She asked them to share how they got the answer and also asked them the basis for doing a particular step. Students, however, responded mostly with rules which they had already learnt in tuitions and failing to engage with contexts or conceptual discussion that Swati was trying to initiate. Swati, at times, reverted to procedural explanation in these situations and at other times tried to justify why it is important to understand why rules work.

> Student: How do we know that it is [the answer]? Suppose we have an example, some sums. How do we solve [without rules]?

> Teacher Swati: Slowly and slowly we will form the rules. These we are just seeing why it happens, then we generalise and get the rules. Rules you have to use, otherwise every time you will not have a number line... but what is applicable for small numbers will be applicable to big numbers. (Class room excerpt, 27-10-10)

In the first phase, the teacher's response to students' answer in Initiation-Response-Evaluation (Mehan 1979) pattern of interaction was either evaluation or explanation of steps or procedure followed by next question for the next step. However, Swati used a variety of ways to respond to students in the second phase like she asked students to evaluate answers given by a student, solicited other answers or methods, and compared different solutions/answers that were shared in the lesson, asking student to give reason for the step, not accepting answer based on rule and asking for explanation using context or model, discussing wrong answers and having extended discussion on one problem or an answer. While the pace of Swati's lesson was established by response of 'bright' students in the first phase, she was aware of the authority that these students have in shaping the response of other students and deliberately included students who had been silent and asking students to think on their own rather than base their answer on somebody else's response. However, Swati would have extended discussions on one problem or answer in the second phase and at other times would revert to procedural explanation.

Thus in the second phase, we see the teacher exercising her agency to make conscious decisions to address the challenges in classroom which were impeding development of reasoning by adopting practices like using contexts and models for developing explanation of procedures, using context-based tasks and constructing tasks on her own, asking more why questions, expecting students to give explanations and evaluate answers and comparing multiple answers/solutions.

10.2.5.2 Affordances and Constraints for Shifts in Practice

The tensions among the beliefs held by Swati between focus on transmission of procedures or developing student reasoning, lack of knowledge to support focus on reasoning in teaching, textbook based pedagogy and assessment, and the culture of doing topics ahead of class through tuitions constrained Swati in shifting her practices to reasoning based teaching practices. However, reflection on her own practice and beliefs during workshops and post-class discussion with researcher contributed to acknowledgement of tensions, discussion on ways to resolve these tensions and acknowledgement of change in beliefs and practice. Swati believed that the workshop in the second phase also contributed to enhancing her knowledge of meanings of integer and operations and contexts where integers are used, which she found useful in her practice.

As discussed, the classroom observations in the first phase indicated the teaching to be procedure focused suggesting that the teacher valued procedures in mathematics. However, in the first interview with Swati prior to the classroom observations had indicated that although she viewed mathematics as more than just calculations she was uncertain about the depth to which mathematics can be discussed in classrooms. She reported that she addressed students' wrong answers by explaining the procedure again and that she asks weak students to practice a particularly easy set of problems, again and again, indicating her belief that students are not capable of engaging in conceptual discussion.

> Math is not just calculation… like when you go in higher classes, calculation doesn't come into picture at all… You start with calculation. But then that's not the end. That is just the beginning (…) too much depth confuses children, but may be it leads to conception [conceptual understanding]. (Interview excerpt, 25-05-09)

At several points in the interview, the tensions between procedural focus and reasoning were evident. On the one hand she acknowledged that teaching should be about "making the concepts clear like why we are doing this, why is it so," while on the other hand she described a good student as one who would be "writing all the steps" clearly and neatly as an indication of "reasoning it out". She justified focusing on practice for "writing all the steps without which he(sic) might lose marks in exam". She asserted that for learning mathematics, practice is essential and at least 50 problems need to be done "because in one go if they solve problems then only they will get [i.e. understand]".

The lack of conceptual and reasoning focus in Swati's teaching could have been due to limited knowledge of concepts or central ideas of mathematics and pedagogical content knowledge. In the interview, she said that she was not sure about dealing with conceptual difficulties of students when they make an error— "if a student did not understand after explaining second time", she said, "I stop there…. May be some other concept is involved." In her teaching also she was not able to give an adequate conceptual explanation, which would be accessible to students and would often revert back to procedural explanation citing it as being "easy" for students. For example, she explained how to convert "a subtraction problem to addition problem" by telling which signs to change and did not choose to discuss the meaning of integers as increase (positive) and decrease (negative) (i.e. substituting with additive inverse). However, she appreciated the value of the concept of 'additive inverse' as she felt that "whatever method we use for subtraction we are using additive inverse" indicating that she is talking about the underlying structure in all models and contexts for illustrating subtraction of an integer.

The workshops in second phase helped in developing knowledge of meaning and contexts for integers and their operations, which helped Swati in identifying the conceptual challenges faced by students in learning integers like "just like 3 is a whole number, −3 is also a number." She concluded that "first it is important for us to understand… these 3 senses [integers in different contexts, be used to represent state, change or relation] makes us more clear. This is what happened with us... we learnt it here only and it made our concept of integer more clear so it is better in teaching."

One of the constraints in shifting towards reasoning based practices was the use of textbook based pedagogy and assessment. In Phase 1, for Swati, her role as a teacher was to complete the exercises given in the textbook as a way of covering the syllabus. Assessment questions were also given from the textbook or were similar to the ones in the textbook. Since many students had already done the textbook exercises ahead of class they knew the rules to get answers and were thus reluctant to engage in conceptual discussion. Swati cited students valuing of procedures and rules as the reason for the lack of talk based on reasoning in her class. Later, however, she moved to realising her own role as a teacher in propagating this norm and lack of knowledge to support reasoning by doing activities.

> Swati: They[students] just want a rule. What you don't know put in the rule and find out.
>
> Researcher: It is a problem that the way maths has been dealt with throughout their lives.
>
> Swati: Throughout their- throughout our lives. At least now that change has taken place like these activities. (...) More is with us than with them. Because we are teaching them so. Mostly with us it is like that only... because of these workshops change is coming. (Interview excerpt, 5-9-9)

In another instance, she attributed the lack of reasoning in students to the pace of instruction. She realised that the locus of the problem of students' attitude is her own teaching rather than students' disinterest in knowing reasons. She acknowledged the need to change one's own teaching in her session for peer teachers at the end of phase 2.

> ... I know it is going to take a very long time. You know first we have to change ourselves (....) We also want to do things quickly. We are worried about the portion [completing the syllabus] so we want to do fast.... (Workshop excerpt, 26-11-10)

Swati used to consider the students who respond quickly as 'bright' students. Students' performance in exams was the goal influencing use of practices like emphasis on specific questions, speed of solving and differential treatment to weak students so that they pass the exams. In the second phase, the teacher and researcher collaboratively developed a test after looking at several books for conceptually based questions. After seeing student performance, Swati deconstructed the term 'bright' student for herself and shared this as an important insight with other teachers in the workshop.

> Students whom we call bright are not really bright because it's just that they have already done the chapter and thus know the answers but if we twist the question they are not able to answer. They don't know the basics but they will solve it. (Workshop excerpt, 8-09-10)

Thus focus on solving textbook questions was inhibiting Swati in really engaging with student understanding. Her realisation (reflected in the excerpt above) and questioning of her beliefs made her open to using real life contexts and questions beyond the textbook while focusing on students' understanding. She shared the motivation for bringing this change in her practice in the session for her peers.

> … some children had already done. They knew the answer but when I asked them to explain they were not able to explain so taking such an activity made them also think. When I asked the reasoning they told Madam we don't know the reason but we know this is the answer. Because they have learnt the rules. They directly learnt it…. Learning rules is easier and those bright children you know they are able to learn the rules very fast. What is the disadvantage [is that] they don't want to know the reason (…) so that is advantage of having something different in the class which is not there in the textbook. (Workshop excerpt, 26-11-10)

The experience of participating in the study helped her to reflect and change her practice and beliefs which she acknowledged at various instances during the study like she proposed that rules should be taught in the end as once students know rules they do not want to engage in understanding how one can arrive at rules.

> Rules should come in the end because once they learn the rules they stop, you know, to verify whether in reality, it is true or not, like how it has come. They don't want to know also how it has come, they just blindly apply that (Interview excerpt, 11-11-10)

While discussing with the researcher about student responses, Swati discussed how learning mathematics by just practising without understanding impedes their ability for reasoning, which is a modification of her earlier belief that a lot of practice helps in learning.

> What I felt you know, by practice they do it- it is not by- more of practice and less of understanding, that is how they are doing maths (Interview excerpt - 11-11-10)

> I think explaining [to] them and understanding is more important than, you know, than just practicing. So unless both are done together learning won't be proper. (Workshop excerpt, 26-11-10)

In phase 2, Swati and other teachers were engaged in developing resources for teaching of integers (tasks, contexts and activities). She reflected on the value of this exercise.

> Actually, we did it in so much detail here so, I could- I was more aware I realised that the students need a clearer understanding of integer, otherwise, we would clearly say 'No, not like this- do like this'. This is how we used to deal, so that is the change in us I could observe. (Workshop excerpt, 20-11-10)

> Earlier when I used to teach, I used to explain to students and then wait. If they did not get it, I used to explain, again and again, the same thing. But now, I have realized that we need to do things differently. (Interview Excerpt, 25-11-10)

In the process of reflecting on her practice Swati became aware of how rules are valued more by students rather than understanding or reasoning and her role in supporting this through her teaching. Textbook based pedagogy and assessment reinforced learning of procedures and Swati initially considered success in solving textbook questions to be an indication of understanding. In the light of students' performance on questions not in the textbook, the teacher made a distinction between solving by knowing the procedure and understanding. This insight made her look afresh at how she is assessing students' understanding and categorising bright students. While Swati's teaching was focused on teaching of rules and procedure, her reflections in the second phase indicate her realization that knowing rules impedes development of understanding making her revise her view that a lot of practice was necessary for learning of mathematics.

10.3 Discussion and Conclusion

The findings indicate that Swati experienced tensions in her beliefs concerning procedural versus reasoning focus in her teaching, possibly precipitated by the implementation of curriculum reforms that emphasised development of conceptual understanding among students. The curriculum places emphasis on students making sense of mathematics by "engaging every student with a sense of success, while at the same time offering conceptual challenges to the emerging mathematician" (NCERT 2006). However, Swati's classroom practice was largely focused on procedure with the goal of developing procedural fluency. Reflecting with the researcher after classes helped her in resolving these tensions and developing a conviction that reasoning and understanding are as, if not more, important for mathematics learning than practising procedures. It also helped in identifying the challenges and constraints in bringing about intended changes and thinking of ways to move her practice towards a focus on reasoning. However, Swati exercised her agency in the second phase to actually explore practices for reasoning which were supported by the insight that she got from designing assessment beyond the textbook and engaging in development of resources including meanings, models, contexts, tasks and activities in the topic study workshops on integers. Participation in the workshops and collaboration with the researcher in school helped the teacher in developing alternative images of teaching from the commonly prevalent practice of teaching mathematics, which is teacher-centred and focused on procedures. The conscious decisions to explore practice in Phase 2 indicated that Swati is developing an identity of a teacher who focuses on reasoning in her lessons.

We have presented an analysis of 4 lessons of all the lessons observed for teacher Swati and an in-depth analysis of teaching in both the phases might illuminate further about the nature of change in practice. We chose these 4 classes as two ends of the journey of the teacher, representing the starting point and the ending point. The practice in the first phase was fairly consistent but not so in the second phase and thus these 2 lessons in the second phase may not adequately represent practice in second phase but serve the purpose of providing an example of types of practices Swati was exploring.

The findings indicate how teacher 'change' is actually a continuous process of professional growth (Clarke 1994) which involves teacher in negotiating the beliefs she holds and practices that she prefers. Further, it places the teacher in the position of negotiating the social norms in terms of rejecting the textbook based pedagogy and focusing on reasoning when students and the larger societal culture supports learning of procedures and focus on answers rather than understanding. Teachers' professional growth, although driven by the agency of the teacher, is influenced by participation in different professional development contexts that allow teachers to re-negotiate their identities and support exploration of practices. Swati's case is thus interesting as it represents the 'teacher-in-flux' as a result of curriculum reform context in the country and helps to see practice as well as beliefs undergoing a continuous process of negotiation in a social context of participation in workshops with peers and collaborative reflections with researcher rather than being interpreted as static entities. This study thus indicates the potential for developing professional learning communities comprising teachers, teacher educators and researchers to engage in the enterprise of analysing and developing teaching of mathematics in schools.

References

Brodie, K. (2012). Professional learning communities and teacher change. Paper *presented at 12th International Congress of Mathematical education*. Seoul, Korea: ICME.

Clarke, D. (1994). Ten key principles from research for the professional development of mathematics teachers. In D. B. Aichele & A. F. Coxford (Eds.), *Professional development for teachers of mathematics (pp37-48)* (pp. 37–48). Reston, VA: National Council of Teachers of Mathematics.

Clarke, P. (2001). *Teaching & learning: The culture of pedagogy*. New Delhi: Sage Publications.

Clarke, D., & Hollingsworth, H. (2002). Elaborating a model of teacher professional growth. *Teaching and teacher education, 18*(8), 947–967.

Goldenberg, C., & Gallimore, R. (1991). Changing teaching takes more than a one-shot workshop. *Educational Leadership, 49*(3), 69–72.

Gresalfi, M. S., & Cobb, P. (2011). Negotiating identities for mathematics teaching in the context of professional development. *Journal for Research in Mathematics Education, 42*(3), 270–304.

Guskey, T. R. (2000). *Evaluating professional development*. Thousand Oaks, CA: Corwin Press.

Hill, H. C., Blunk, M. L., Charalambous, C. Y., Lewis, J. M., Phelps, G. C., Sleep, L., et al. (2008). Mathematical knowledge for teaching and the mathematical quality of instruction: An exploratory study. *Cognition and Instruction, 26*(4), 430–511.

Jaworski, B. (2008). Building and sustaining inquiry communities in mathematics teaching development: Teachers and didacticians in collaboration. In K. Krainer (Volume Ed.) & T. Wood (Series Ed.), *International handbook of mathematics teacher education* (Vol. 3). *Participants in mathematics teacher education: Individuals, teams, communities and networks* (pp. 309–330). Rotterdam, The Netherlands: Sense Publishers.

Kumar, R., Subramaniam, K., & Naik, S. (2015). Professional development workshops for in-service mathematics teachers in India. In B. Sriraman, J. Cai, K. H. Lee, F. Lianghuo, Y. Shimizu, C. S. Lim, & K. Subramaniam (Eds.), *The first sourcebook on Asian research in mathematics education: China, Korea, Singapore, Japan, Malyasia and India* (Vol. 2, pp. 1631–1654). Charlotte, NC: Infoage Publishers.

Matos, J. F., Powell, A., Sztajn, P., Ejersbø, L., Hovermill, J., & Matos, J. F. (2009). Mathematics teachers' professional development: Processes of learning in and from practice. In *The professional education and development of teachers of mathematics* (pp. 167–183). Springer US.

Mehan, H. (1979). *Learning lessons: Social organization in the classroom* (p. 80). Cambridge, MA: Harvard University Press.

Miles, M. B., & Huberman, A. M. (1994). *Qualitative data analysis: An expanded sourcebook*. Sage.

National Council for Educational Research and Training. (2006). *National focus group on teaching of Mathematics Report*. New Delhi: NCERT.

National Council for Teacher Education. (2009). *National curriculum framework for teacher education: Towards preparing professional and humane teacher*. New Delhi: NCTE.

Pajares, M. F. (1992). Teachers' beliefs and educational research: Cleaning up a messy construct. *Review of Educational Research, 62*(3), 307–332.

Putnam, R. T., & Borko, H. (2000). What do new views of knowledge and thinking have to say about research on teacher learning? *Educational researcher*, 4–15.

Richardson, V., & Anders, P. (1994). The study of teacher change. In V. Richardson (Ed.), *A theory of teacher change and the practice of staff development: A case in reading instruction* (pp. 159–180). New York: Teachers College Press.

Saunders, W., Goldenberg, C., & Hamann, J. (1992). Instructional conversations beget instructional conversations. *Teaching and Teacher Education, 8*, 199–218.

Sparks, D., & Loucks-Horsley, S. (1990). Models of staff development. In W. R. Houston (Ed.), *Handbook of research on teacher education* (pp. 234–250). New York: Macmillan.

Stein, M. K., Grover, B. W., & Henningsen, M. A. (1996). Building student capacity for mathematical thinking and reasoning: An analysis of mathematical tasks used in reform classrooms. *American Educational Research Journal, 33*(2), 455–488.

Swan, M. (2006). *Collaborative learning in mathematics. A Challenge to our Beliefs*. Leicester, UK: National Institute of Adult Continuing Education.

Vescio, V., Ross, D., & Adams, A. (2008). A review of research on the impact of professional learning communities on teaching practice and student learning. *Teaching and Teacher Education, 24*(1), 80–91.

Wenger, E. (1999). *Communities of practice: Learning, meaning, and identity (learning in doing: social, cognitive and computational perspectives)*. Cambridge: Cambridge University Press.

Chapter 11
A Community-Based, Practice-Oriented Teacher Professional Development Program: Changes in Teaching Culture in Korea

Oh Nam Kwon, Jee Hyun Park, Jung Sook Park and Jaehee Park

Abstract This chapter introduces a conceptual framework and practices yielded by research into a teacher Professional Development program focusing on teacher community for mathematics teachers to increase professionalism. Conceptually, it was distinguished from the other training programs in terms of the participants, curriculum, and methods. The teacher communities consisting of 3 or 4 teachers from the same school, as well as mentor and sub-mentor, master or professional teachers with professional expertise and executive capability. There were a total of 28 teachers from 9 schools and 18 mentors and sub-mentors supported each school. The curriculum of our program includes some process practicing and reflecting of teachers' communities on their own classes. The program's structure required active participation. Through our program, the teachers improved their teaching competency. Also, the operational ability of the teacher learning communities was improved. A teaching and learning community culture had been formed in each school, which showed that the community could be voluntarily operated although our program is over. The teachers volunteered to open their classes to teaching community members, rather than avoiding it, breaking awareness that open class regarded as the place of evaluating them until now in Korean classroom culture. The opening classes became a new method to improve teaching competency of community.

O.N. Kwon (✉)
Seoul National University, Seoul, South Korea
e-mail: onkwon@snu.ac.krQuery

J.H. Park
Banpo High School, Seoul, South Korea
e-mail: jeannei@sen.go.kr

J.S. Park
Yangjae High School and Hanyang University, Seoul, South Korea
e-mail: pjs1924@sen.go.kr

J. Park
Gyeonggi Science High School for the Gifted, Suwon, South Korea
e-mail: udmji@snu.ac.kr

B. Kaur et al. (eds.), *Professional Development of Mathematics Teachers*, Mathematics Education – An Asian Perspective,
DOI 10.1007/978-981-10-2598-3_11

Keywords Professional development · Teacher education · Teaching practice

11.1 Introduction

To date, the programs to develop teachers' professionalism have been based on the expectation that if teachers acquire knowledge related to their classes, teaching quality will automatically improve. However, many research studies have shown that simple transmission of knowledge is not effective in changing the beliefs, attitudes, or teaching practices of teachers, all of which directly impact teaching quality (Shulman and Shulman 2004).

Teaching by a professional teacher is basically practical work, that is, practical teaching knowledge comes through experience, as well as the teachers play a central role in generating knowledge of practice (Cochran-Smith and Lytle 1993). Increasing the teachers' professionalism cannot be achieved merely by fostering an in-depth understanding of the content or by arranging hands-on lectures by external experts. But it must be based on the actual class or context setting faced by the teachers in question, with the direct participation of those teachers in defining their authentic context and difficulties in implementation of knowledge or concepts (Cobb and Bowers 1999; Chapman 2012). Accordingly, recently teacher training has been conducted in a way that helps teachers develop tasks or activities they can use in actual classes in connection with their context of practice.

In Korea, diverse research on mathematics teacher training program has been conducted to create a change in this regard. But so far, these studies have mostly focused on theoretical analysis and policy suggestions, far from training practices (Kwon et al. 2012; Lee and Jang 2012). In addition, even studies on in-service teachers and teacher training program have been mostly one-time studies that focus on individual changes in teachers in short-term training programs (Choi 2013). These researches suggested that to develop an effective Korean professional development system (PDS), theories and practices for teaching should be integrated using various programs; and there is a need for an effort to develop a training program that ensures continuous development of teachers' professionalism.

Meanwhile, according to Goldsmith et al. (2014), 36 of 106 articles published from 1985 to 2008 were related to "teachers' collaboration/community." This shows there has been a reasonable amount of interest in teacher community in research on teacher PD programs. Whether it is cooperation among colleagues with horizontal relationships or coaching and mentoring relationships, collaboration among teachers enables them to share their experiences in class implementation, helps them make instructional decisions, and broadens their understanding of the student learning process (Chazan et al. 1998; Kazemi and Franke 2004). In turn, community between teachers working in similar school atmospheres or learning environments, mutual affinity between teachers, and jointly agreed-upon objectives for teaching activities can all foster mutually beneficial relationships among teachers (Kim et al. 2010; Lee et al. 2013). Given the wide recognition of these benefits, interest in teacher community is on the rise, and if training is conducted to foster teacher community within individual schools, the professional development of math teachers will be improved.

This chapter will introduce a conceptual framework and practices of a teacher PD program focusing on teacher community for elementary, secondary school math teachers. The program was yielded by research (Kwon et al. 2014) to increase professionalism, where professionalism of mathematics teachers is regarded as a factor enhancing their ability to improve their lessons and help students' learning, and hope them cope with changes related to the implementation of a new curriculum in Korea; the study was conducted for 5 months, from September 2013 to January 2014. The teacher PD program is not intended simply to foster passive participation by teachers in a prepared training program, but instead to help them become actively engaged in their teaching practice in the context of a professional community, and to encourage them to become practical teacher researchers themselves.

11.2 Theoretical Foundation of the Practice-Oriented PD Model

Kwon et al. (2014) found that through math teacher PD program rooted in the school community, teachers' understanding of the "storytelling" and educational themes of "science, technology, engineering, arts, and mathematics (STEAM)" was enriched and their ability to implement those themes in class is improved. The PD program discussed here was intended to foster teachers who could continue to enhance their professional capabilities as practitioners and researchers, on the basis of the following three frameworks.

11.2.1 Curriculum for Teacher's Practice

Existing teacher training programs have largely focused on in-depth subject knowledge. In contrast, this PD program presented here stresses teaching practice to help teachers identify relevant issues for current math education and apply those issues, such as storytelling, STEAM, in actual classes.

The program was based on "Situated cognition and learning" theory, which began to be discussed by many researchers in the late 1990s (Anderson et al. 1997; Greeno et al. 1997). The core idea of situated cognition and learning is that knowledge, thinking, and learning should be understood from the situational perspective (Greeno et al. 1996). Lave and Wenger (1991) emphasized the importance of context and situation for explaining the thoughts of human beings, and stressed the importance of understanding that how we think and what we think is related to the situation where it takes places.

Lave (1996) focused identifying the influence of social interaction on individual cognitive changes. The core assumption of the theory is that knowledge is situated in context, context is linked with activities and culture, and learning requires an integrated understanding of these three elements as well as the target knowledge (Brown et al. 1989). That is, they viewed learning as the interaction of the context, culture, and activities in and through which the learning is done. In this view, how teachers learn teaching methods is not different from how students study (Putnam and Borko 2000). Teacher training programs that focus only on providing information and remembering knowledge often do not directly help teachers solve problems stemming from the various contexts they may face in actual classrooms. Thus, a key issue in teacher training is how we can create learning experiences that can help teachers apply their knowledge in class. Situated cognition and learning theory suggests that the classroom situation is directly provided to teachers, and that if teachers' knowledge and learning is really situational, the most effective teacher training should focus on actual class implementation, including the provision of knowledge and contextual activities based on this situation. Ultimately, both theory and practice need to be considered.

Accordingly, the PD program to be presented in this study divided class teaching into four components—tasks, teaching instruments, discourse, and assessment—and attempts to provide teachers with learning experience in the actual context of their own school classrooms (Fig. 11.1). The curriculum used was formed to help

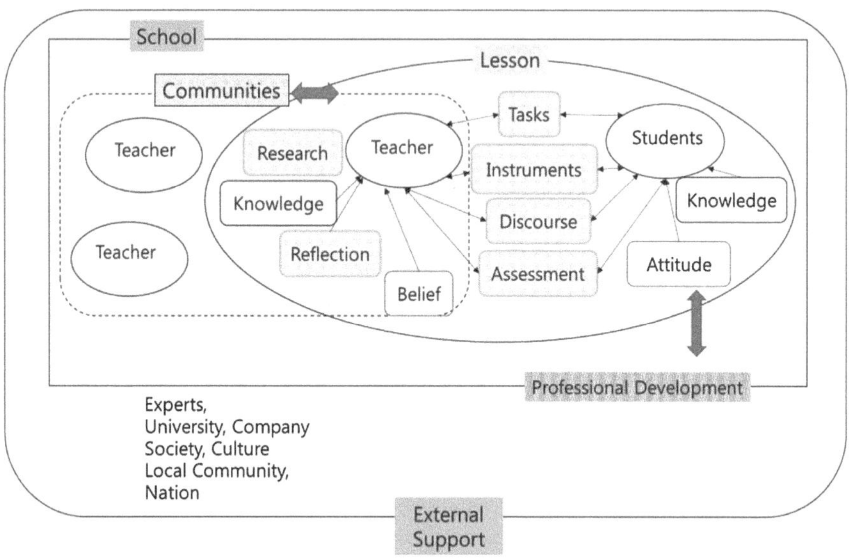

Fig. 11.1 Factors and interactions in the lessons and the lesson environment

teachers play the role of practitioner researchers through the process of implementation and reflection.

The specific themes pursued are mathematics lessons using storytelling for elementary students, which have been gaining attention recently in mathematics education in Korea, and STEAM-based mathematics lessons for middle and high schools. The term storytelling is a compound word of story and telling, so storytelling means making a story and telling a story to others. The 2009 revised curriculum reported that storytelling should be applied to the introduction of textbook in elementary school. The purposes of storytelling are to ensure students' interest in mathematics and to foster students' imagination and creativity. Figure 11.2 shows the introduction of elementary textbook reflected storytelling. The unit used the story of "The Emperor's New Clothes" (Kang 2013, pp. 126–129 and 142–145) to explain the need of unit of measure.

In addition, STEAM (Science, Technology, Engineering, Arts, and Mathematics) education has been emphasized as a keyword in the field of education. Particularly, the purpose of STEAM curriculum focused on mathematics is to help students make use of mathematical ideas to analyze, interpret, and reinvent various natural phenomena, social phenomena, and art through activities that develop creativity, problem-solving ability, and artistic sensibility. Since the 2009 revised curriculum, various STEAM materials has been developed and applied to the middle and high schools. Figure 11.3 shows that the examples from high school mathematics textbooks (Kim 2014a, p. 100; Kim 2014b, p. 196), which used the

Fig. 11.2 An example of the storytelling in elementary textbook (Reproduced with permission from Kang 2013, pp. 126–129 and 142–145)

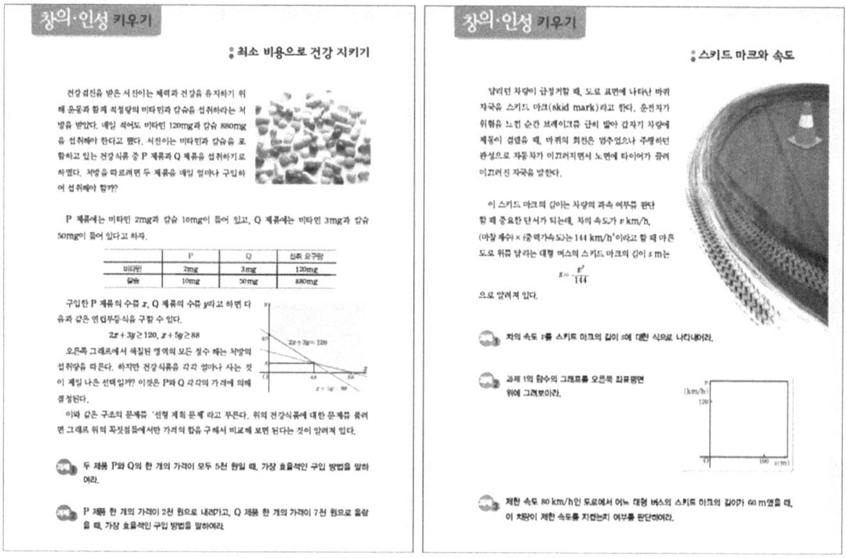

Fig. 11.3 Examples of STEAM-based mathematics in high school textbook (Reproduced with permission from Kim 2014a, p. 100; Kim 2014b, p. 196)

connection of art and mathematics and the connection of technology, mathematics, and real life.

The contents of the teachers PD program were developed specifically to convey these themes in the classrooms and to solve problems teachers encounter with the transmission of these themes in classrooms. With this goal in mind, the contents of the PD program was built to help teachers experience and practice tasks, discourse methods, and reflection, and to present theories related to each theme such that teachers can change and apply it as needed in their own classroom situations. Thus, different themes were assigned to elementary students and to middle and high school students to reflect their different situations, and opportunities were provided for the teachers to experience and carry out task-setting; discourse methods to be used to conduct classes; reflection, which forms the mechanism for changes in class practice, and with them to transmit relevant theories relating to each theme.

11.2.2 Multitier Teacher Community of Practice

In many teacher training programs, individual teachers apply for and receive training individually. In contrast, the present program provided both researchers and participants with the opportunity to be closely associated with each other during the whole process of the program.

Wenger (1998) defined the community of practice as an informal gathering of people sharing common tasks. The community of practice approach has drawn attention as a method of developing teachers' professionalism in order to change educational practices in schools, because many teachers find it very difficult to incorporate what they have learned during lectures in training programs into classrooms (Greenberg and Baron 2000; Shulman and Shulman 2004). Moreover, teachers may not receive proper feedback on their lessons, and may be left to cultivate their professionalism through personal trial and error in the actual classroom, isolated, without communication among colleagues, and trapped in the individualistic culture of teachers (Hargreaves 2000). In our program, in contrast, teachers' learning is based on the premise of mutual disclosure of knowledge and practices, carrying out common tasks together, and working together throughout the whole process of planning, teaching, and evaluation. Individual teachers-as-learners are members of various discourse communities, who start as peripheral participants in a professional community and later on come to play a leading role by actively participating in the culture. Lave and Wenger (1991) referred to this learning process as legitimate peripheral participation, in which learning leads to increased participation through interaction with other members of the community, and in which, through this situated learning process, learners can establish their identity and competence as members of the community. In the process of helping legitimate peripheral participants enhance their professionalism and grow into full participants, collaborative learning should be emphasized.

As teachers working at the same school teach students in the same context, often the same students, it is easy to discuss the construction and practice of lessons and to share the same educational goals (Seo 2013). Since this approach was considered important for our program, this PD program participants were limited to mathematics teachers serving the same school. The "community" here was intended to consist of the PD program participants at a single school, as well as mentors, master, or professional teachers with professional expertise and executive capability. Sub-mentors were also included members of communities as the practical assistance for the research. Mentors and sub-mentors acted as catalysts, helping PD program teachers who were still peripheral participants grow to become full participants and implement learning and teaching content and practices effectively. This was not a large-scale study; the number of participating schools and thus communities of practice was limited to four or five (with at least three teachers from each school).

Mentors and sub-mentors discussed how to improve classes with other participating teachers and discussed the overall operation of the program as with the researcher; if any support was necessary, they requested the needed support and provided assistance. Therefore, in a broad sense, we see that the teacher community in a teacher PD program like this one can include not only teachers themselves but also the researchers organizing the PD program and the instructor lecturing in it. Figure 11.4 illustrates a multitier teacher community in a teacher PD program like this one.

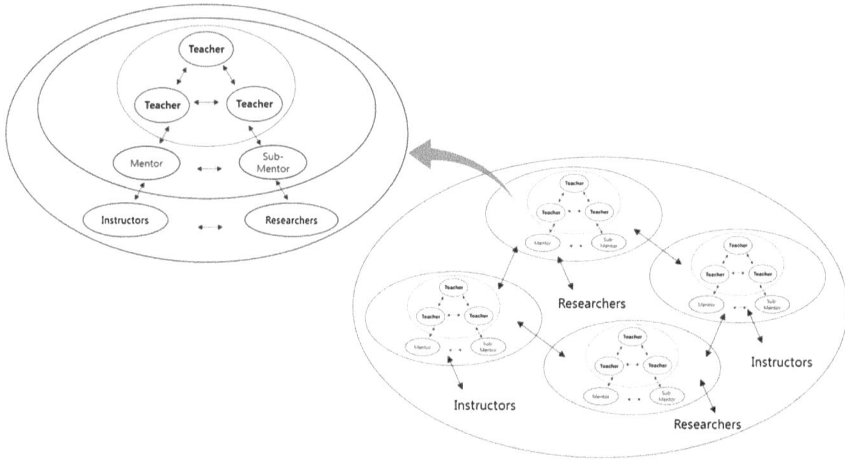

Fig. 11.4 Multitier teacher community

The left-hand circle in Fig. 11.4 describes a school community, while the right-hand circle shows how the school community expands to multiple communities that interact with each other, sharing experiences.

11.2.3 Structure of Participation

This PD program was driven entirely by participants, from application to implementation, in order to ensure their active participation. Participants submitted applications that described what they wished to learn and what their goals were for the PD program, and sat for interviews with the researchers to determine their participation in the program. They attended lectures, discussions, workshops, and planning lessons appropriate to the situation of each school. In class, participants were involved in diverse activities such as lesson planning, teaching implementation, and discussion after class. In particular, for each stage, reflection activity was mandatory. The reflection activity can be used in one of the effective way in teacher education. Schön (1987) discusses reflective practice and argues that education theories should be accompanied by practices based on them. According to Zeichner (2001), the international movement in teacher education and reflection can be seen as a reaction against the view of teachers as technicians who merely apply formulas dictated by others who are, more often than not, removed from the classroom. Teachers should strive to be reflective practitioners, capable of analyzing their own practice. Teachers could continuously develop their professionalism through reflections on their practices.

Participants went beyond the use of the theories and skills they learned during the intensive education segment of the program to build new knowledge in response to

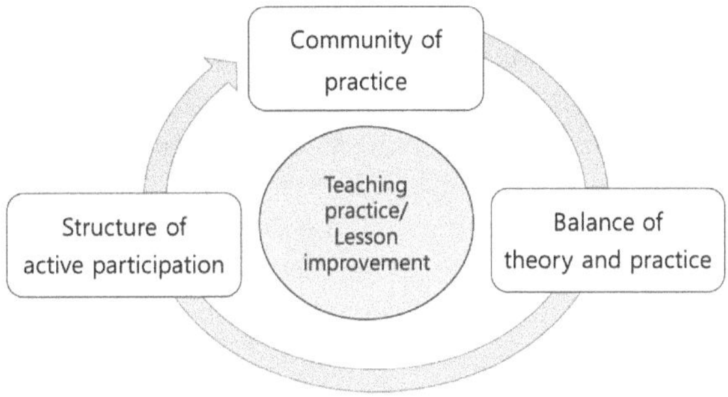

Fig. 11.5 Conceptual framework for the PD program (Kwon et al. 2014)

new classroom situations. In particular, such knowledge was formed while teachers planned and executed lessons and discussed them with their community of practice —not only a form of the "reflection-in-act" but also "reflection-on-acts" mentioned by Schön (1987). This process encouraged teachers to become practical researchers, that is, practitioners with an eye on theory or theorists with an eye on practice.

In our mathematics teacher PD program, the content, the participants, and the method were all aimed at improvement of actual lessons in school classrooms; in order to achieve these changes, a conceptual model for the PD program (Kwon et al. 2014; Kwon 2015) was developed, as follows Fig. 11.5.

Previous training programs delivered teaching knowledge to individuals through lectures, and classes seemed to have been performed based on this acquired knowledge; however, such knowledge has not led to any significant improvement to lessons (Greenberg and Baron 2000; Shulman and Shulman 2004). This model was intended to overcome that limitation.

11.3 Operation of Community-Based Mathematics Teacher PD Program

The teacher PD program developed by this research emphasizes the process of building classes together with mentors, with active participation of each teacher who is a member of the community of practice. Thus, the program was named "Professional Development Program Placing Teachers Together." It was divided into three stages. Twelve teachers from four elementary schools and sixteen teachers from five high schools participated in the program, and eighteen mentors and sub-mentors drawn from the nine school communities assisted the progress of the program.

PD program participation was community based, and focused on combined professional development in theories and practices; active participation of participants was encouraged. The effects of the program were analyzed and considered from various perspectives. Class plans, video clips of classes, and minutes of community of practice meetings were collected to consider the cases of the various participants. Qualitative material such as mentor journals and interviews were also gathered. In addition, quantitative analysis of the program was done using surveys before and after the PD program. Kwon et al. (2014) reported that the surveys showed increased theoretical knowledge and understanding of class practices related to "storytelling math classes" and "math-focused STEAM classes" among the participants. In addition, it was found that participants had become aware of the benefits of the relatively long-term community-based PD program that this program aimed to be. These changes were also observed in their overall course practices, during the process of program, and in their reflections.

11.3.1 First Stage: Preparatory Intensive Course

As the first stage in the program, preparatory course was implemented. This stage consisted of attending an intensive training in how to adapt the program material for school classroom situations. Although the themes of PD program were different for elementary school and high school teachers, they attended the same lectures and worked together on common tasks such as reflection, operation of the community of practice, and engaging in teacher research activity. The program was run every saturday for 3 weeks from September 28, 2013 for 18 h in all (6 h per day). Teachers were expected to improve their lessons by incorporating what they had learned. To give them a chance to do this, we set an interval of 1 week. Major training objectives and content during the 3 weeks were as in Fig. 11.6.

In the training on storytelling for elementary school teachers, topics of lectures included the meaning of storytelling; comparison of storytelling textbooks and general textbooks; development of storytelling tasks; writing of a class guide for storytelling mathematics class; analysis of an elementary school class video clip showing a storytelling lesson; and preparation of a class plan imagining students' response. In the "mathematics-focused convergence class" training for high school math teachers, the topics were meaning, ideas, and nature of the math-focused convergence class; an example of such a class; experiential activities utilizing 3D printing; task planning, and class operation; analysis of teachers' discourse on this class; and preparation of a class focusing on imagining students' response.

One of the major features of the first stage is that training was not conducted entirely by lecture but also by activities and workshops, so that teachers could experience the target practices, followed by a summary lecture at the end. Last,

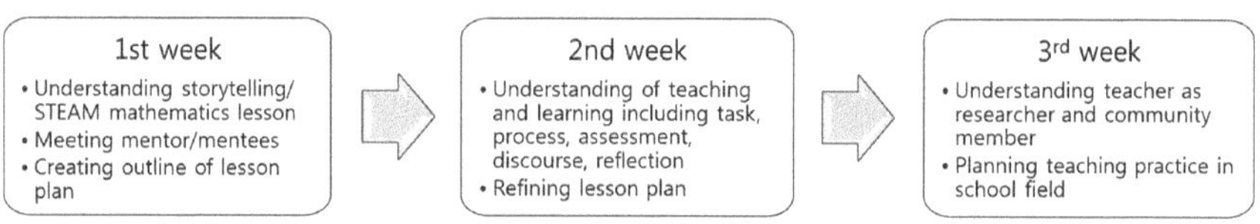

Fig. 11.6 Goals and contents of first stage: preparatory intensive course

participants were guided to establish detailed lesson plans that would suit their actual classrooms.

Throughout the process, the researchers reviewed the training progress everyday and analyzed teachers' needs regarding the training program so that the content of training and tasks could be modified to meet teachers' needs. Mentors and sub-mentors also participated in training, sharing time and effort and building relationships with teachers, and assisted them to prepare to apply what they had learned to actual classrooms. Thus, in the first stage, teacher communities were guided to share visions and establish concrete goals to move them toward adaptation to real-world situations.

11.3.2 Second Stage: Teaching Practice and Collaboration

Training to adapt the learning conducted in the program to school classrooms was implemented over 2 months. During this period, participants of our PD program operated a teachers' community at their own schools and established lesson plans reflecting the themes of the training program, utilizing what they had learned in the preparatory course, in preparation for visits to their classrooms by the facilitators. In this stage, participants were allowed to communicate online with mentors and sub-mentors via e-mail or social networking sites to discuss preparation of lesson guides.

The second stage had three steps: "lesson sharing," "lesson caring," and "lesson nurturing," involving community of practice meetings, online discussions with mentors and sub-mentors, open classes, and site visits by mentors and sub-mentors. In lesson sharing, ideas were freely exchanged to develop teaching and learning plans through regular teachers' community meetings and online discussions with mentors and sub-mentors. Next, in lesson caring, mentors and sub-mentors visited school classrooms, where they watched the participants' classes together with members of the community. Based on the results of visit, specifically of reflection on practice after class, feedback from mentors, and discussion between the teacher

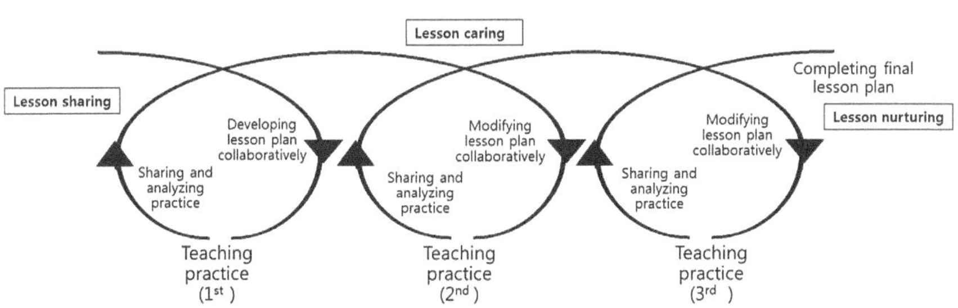

Fig. 11.7 Progress of teacher–community and mentor collaboration

and mentor, teachers improved their teaching and learning plans and applied the improved plans to their classes, repeating the class planning and practice process. Last, in lesson nurturing, teaching and learning plans were finalized by incorporating the discussion that took place during the teachers' community meetings and the site visits. The second stage is illustrated in Fig. 11.7.

Each community repeated the process of applying and caring lessons at least twice. In some cases, after a teacher taught a class, the teachers' community improved the teaching plan and the teacher conducted again in a second class. In another case, after first teaching and meeting of the teachers' community, another teacher practicing lesson another class using by modifying teaching plan. In the other case, the teaching method was improved and then all members of the community taught a class together (team-teaching). All these exercises were done in different ways depending on school situations and characteristics of the community.

Teachers' understanding of the themes was observed in various ways. Although theoretical knowledge is important, it is more important to be able to solve tasks and engage in discourse with students in actual school classrooms and to be prepared for any contingencies. Teachers had already learned theories related to the tasks and evaluation of students' discourse in the classroom during the first stage, and had the opportunity to apply them in actual school classrooms. Based on their experiences during the first stage, teachers went through the cyclical process of discussing their classes in the teachers' community as presented above, reflecting on diverse aspects of teaching in order to incorporate the lessons learned into their next classes.

One elementary school community misunderstood the concept of contextualization (Kwon et al. 2013; Zazkis and Liljedahl 2008) in teaching and learning through storytelling, and thus, at early phase, they introduced a storytelling activity briefly just to attract the attention of the students at the beginning of a class. However, the students interpreted the story and activities as constituting separate elements of the lesson and lost concentration in the later part of a class as a result of being unable to perceive the connections between them; as a result, the teachers felt they needed to use new methods. After discussion in the community, in the next class the teachers tried to increase the use of storytelling in order to inspire students to engage in the activities, using a single story of one hour's duration. All elementary school communities attempted to create storytelling mathematics lessons with a coherent story encompassing the whole chapter, rather than fragmented stories. This was what the training program was basically intended to allow.

In high school classes, which were on the theme of convergence, teachers' understanding of the theme and ability to apply it in the classroom was enhanced significantly. High school communities produced a variety of lessons, depending on the characteristics of the school and grade, on topics including problem posing using technology; integration of mathematics and physics in a social context; measurement by division in solving tax issues in the Joseon Dynasty period (1392–1910); and the use of mathematics in decision-making situation. Teachers, who had not known how to introduce other subjects or real-world contexts into their mathematics classes, succeeded in applying the concept of STEAM in classes using various materials and formats and improving quality of lessons through discussions with the community and mentors and reflecting the response of students.

The use of a cyclical process of collaboration among teachers in communities of practice did not simply produce a change in the content of teaching practice but also a change in the attitudes of teachers. After the PD program ended, many teachers said in interviews that learning is possible in other spaces than schools and that in the process of working together to find solutions to problems, and they felt a sense of solidarity, of working "together." While jointly thinking about and preparing for classes with other teachers in the community of practice, they came to foster trust among their colleagues; and regardless of career experiences and other conditions, they came to recognize the value of others. This was purely the result of genuine collaboration among teachers toward achieving good classes. While preparing for classes based on the spirit of mutual cooperation, teachers felt they were jointly responsible for classes, and they become confident that opening of teaching (open lessons) in classrooms are not for evaluation, but for improvement of the class and gathering of ideas to help it improve. This shows a change of perspective—the opening of classes is not "opening for evaluation" but "sharing for improvement."

11.3.3 Third Stage: Post-Program for Sharing Participation Experiences

The third stage basically consisted of the second round of lecture after adaptation of the learning to school classrooms and training by visit were completed. It took 12 h over 2 consecutive days. Each teachers' community had the opportunity to share and discuss members' teaching experiences. That is, cases of classes where the participants actually applied the lessons learned during the training period were presented, and the mentors and facilitators of the training program analyzed the activities of each community member to derive answers to problems in classrooms. In the third, closing phase of the training program, ways to maintain community among teachers after the end of the training program were sought and discussed.

11.3.4 Procedural Model of the Practice-Oriented PD Program

The three-stage process of PD program can be summarized as in Fig. 11.8.

As mentioned earlier, our PD program consists of preparatory training for adaptation to actual classrooms, on-site training by visit, and post-training. The key procedural characteristic of this research is that in the preparatory intensive training stage, participants were able to acquire concrete, practical content and processes

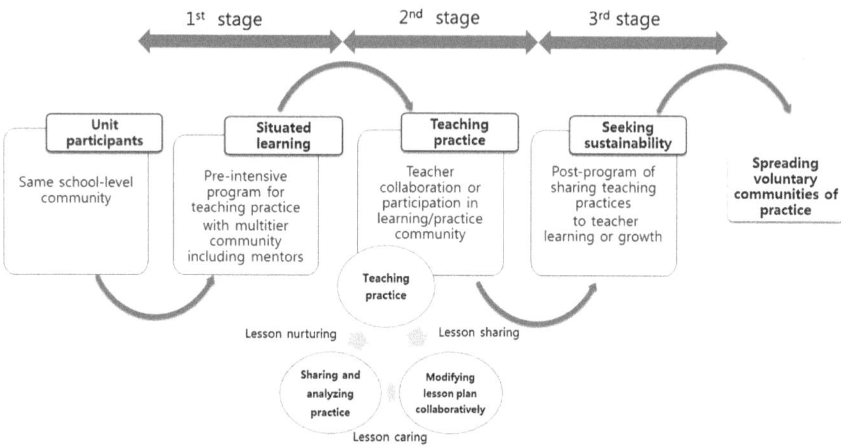

Fig. 11.8 Procedural framework of practice-oriented PD model

that they could apply in the classroom by linking theories with practice, rather than just by receiving one-way theoretical lectures. Moreover, in the on-site training stage in the classroom, which is the most important part of this training program, classes were prepared, implemented, reflected on, and shared within each community, in contrast to previous programs that delegated the burden of applying theories in the classroom to each individual, while the facilitators of the program checked their progress. Experiencing collaboration through a teachers' community, along with the assistance of professional mentors, is a unique feature of this training program.

After the training ended, it was found that teachers obtained more confidence implementing classes on each theme (Kwon et al. 2014). Confidence of running classes on the various themes, at both the elementary school and middle and high school levels, was found to have increased compared with confidence level prior to training. One of the factors contributing to increased confidence about storytelling or the STEAM class themes was the fact that based on theories learned, participants had been effectively guided to create their own class content in their actual school classrooms, maintaining an optimal balance between theories and practices. Other positive factors included the participatory framework available to teachers in planning, implementing, and reflecting on their classes, and the formation of a solid teachers' community of practice with support by mentoring teachers.

11.4 Discussion and Conclusion

The ultimate purpose of the community-based mathematics teachers PD program that was developed by this research is to support continuous development of teachers' professionalism through training, where professionalism of mathematics teachers is regarded as a factor enhancing their ability to improve their lessons and help students' learning. To this end, rather than transferring all responsibilities to individual teachers, their professionalism was enhanced by growth through collaboration and reflection within the teachers' community.

The most conspicuous differences between our PD program and the other programs already are as follows. First, in this program, process is planned and operated with a special focus on the teachers' community, understood as basically a group of teachers serving in the same school, and in a broader sense also including mentors and sub-mentors that support the school community. At least three teachers working at the same school formed one team in this training program, which led to smoother organization of community meetings in terms of time and location, unlike other communities composed of individual members from different schools. In addition, when participants were teaching the same grade of students, school activities and curriculum were almost identical, and thus collaboration was particularly effective and efficient. Furthermore, a smaller teachers' community with only a few members makes it relatively easy to share common goals and directions while coordinating opinions and reaching agreements, making it highly suitable for teachers who have just begun to work with the community of practice model.

Second, our PD model focused on practice with a cyclical procedure. That is, the participants obtained assistance in the practical implementation of the teaching methods and learning content involving community of practice meetings, online discussions, open classes, and site visits by mentors and sub-mentors and going through the cyclical procedure of 'lesson sharing,' 'lesson caring,' and 'lesson nurturing,' This community-based process improved the understanding of the themes by the participants and boosted their confidence as teachers. In addition, the fact that the teachers volunteered to open their classes to teaching community members, rather than avoiding it, provided momentum for the discussion of ways to improve their classes. Since the teachers have regarded their class opened as the place of evaluating them until now in Korean classroom culture, it is the meaningful change. They shared their own individual problems and made them problems of the community, making the community as the effective tool for improving pedagogy. Moving away from the existing classroom culture, where one teacher does not interfere with another's classes, these changes in classroom culture took place as a result of the collaborative construction of classes, and these teachers now open their classes to colleagues comfortably, paving the way for the creation of a new community culture across schools. Participating teachers responded very positively to the question "Were teachers' communities at each school that were run during the training period helpful in improving classroom culture where classes are shared and teachers cooperate with one another?' (Kwon et al. 2014). Moreover, as they reflected on their classes, they came to better understand the role of the community, and even after the end of training, they wanted to preserve the teachers' community at each school and to continue their research into class procedures, as teachers who learn as well as teach.

One elementary teacher participant said during interview after the program ended, "I felt the positive attitude of my colleagues, not attending in the program, toward community activities." Nonparticipating teachers at the same school expressed interest in the activities of the teachers who attended the training program. And at other elementary and high schools, some nonparticipating teachers expressed the intention to join in class improvement activities as members of the teaching community of practice after the PD program. This demonstrates that the need to form a community is shared among teachers who did not participate in the PD program, as well as participants.

The concept and procedural model of the training program developed by this research may be modified to suit the needs of other course subjects than mathematics, so that the model can be applied to the operation of PD programs for these other subjects. This systematic PD program will facilitate sustainable development of teachers' professionalism as teacher researcher, the spread of community among teachers, and the enhancement of teachers' capability to implement the learning material, thereby creating positive change in mathematics education. In fact, inspired by these positive effects, the Korea Foundation for the Advancement of Science and Creativity (KOFAC) is applying our PD program model in its 2014 PD program for elementary school teachers to foster mathematics classes based on storytelling.

References

Anderson, J. R., Reder, L. M., & Simon, H. A. (1997). Situated versus cognitive perspectives: Form versus substance. *Educational Researcher, 26*(1), 18–21.

Brown, J. S., Collins, A., & Duguid, P. (1989). Situated cognition and the culture of learning. *Educational Researcher, 18*(1), 32–42.

Chapman, O. (2012). Challenges in mathematics teacher education. *Journal of Mathematics Teacher Education, 15*, 263–270.

Chazan, D., Ben-Chaim, D., & Gormas, J. (1998). Shared teaching assignments in the service of mathematics reform: Situated professional development. *Teaching and Teacher Education, 14*(7), 687–702.

Choi, J. S. (2013). A study of the effectiveness of a mathematics teacher's training program focused on the construction of test items. *The Journal of Educational Research in Mathematics, 23*(2), 199–212.

Cobb, P., & Bowers, J. (1999). Cognitive and situated learning perspectives in theory and practice. *Educational Researcher, 28*(2), 4–15.

Cochran-Smith, M., & Lytle, S. L. (1993). *Inside outside: Teacher research and knowledge*. NY: Teachers College Press.

Goldsmith, L. T., Doerr, H. M., & Lewis, C. C. (2014). Mathematics teachers' learning—A conceptual framework and synthesis of research. *Journal of Mathematics Teacher Education, 17*, 5–36.

Greenberg, J., & Baron, R. A. (2000). *Behavior in organizations* (7th ed.). Upper Saddle River, NJ: Prentice-Hall.

Greeno, J. G., Collins, A., & Resnick, L. B. (1996). Cognition and learning. In D. Berliner & R. Calfee (Eds.), *Handbook of educational psychology* (pp. 15–46). New York: Macmillan.

Greeno, J. G., & The Middle-School Mathematics through Applications Project Group. (1997). Theories and practices of thinking and learning to think. *American Journal of Education, 106*, 85–126.

Hargreaves, A. (2000). Four ages of professionalism and professional learning. *Teachers and Teaching: Theory and Practice, 6*(2), 151–182.

Kang, W. (2013). *Mathematics 3 (elementary school)*. Korea: Ministry of Education, Science and Technology.

Kazemi, E., & Franke, M. L. (2004). Teacher learning in mathematics: Using student work to promote collective inquiry. *Journal of Mathematics Teacher Education, 7*(3), 203–235.

Kim, C. D. (2014a). *Mathematics 1 (high school)*. Korea: KyoHakSa.

Kim, C. D. (2014b). *Mathematics 2 (high school)*. Korea: KyoHakSa.

Kim, J. H., Choi, W. J., & Shim, J. S. (2010). Narrative inquiry in class criticism: case of community of teacher learning practice. *Secondary Institute of Education, 58*(3), 333–355.

Kwon, N. Y., Lee, E., Park, M., & Park, J. (2012). A study of professional development for mathematics teachers: Based on the investigation of professional development programs in foreign countries. *The Journal of Educational Research in Mathematics, 22*(3), 287–400.

Kwon, O. N., Ju, M. K., Park, J. S., Park, J. H., Oh, H. M., & Cho, H. M. (2013). The study on the development principles for the mathematics textbook based on storytelling and the possibility of implementation. *Journal of the Korean Society of Mathematical Education Series A: The Mathematical Education, 27*(3), 249–266.

Kwon, O. N., Park, J. S., Park, J. H., & Cho, H. M. (2014). Designing and implementing professional development program of multitier teacher community: Focus on "Together making mathematics teacher training program". *Journal of the Korean Society of Mathematical Education Series A: The Mathematics Education, 53*(2), 201–217.

Kwon, O. N. (2015). Creating and sustaining effective professional developments for mathematics teachers in Korea. In C. Vistro Yu (Ed.), *Proceedings of the 7th ICMI-East Asia regional conference on mathematics education* (pp. 17–28). Quezon City: Philippine Council of Mathematics Teacher Educators (MATHTED), Inc.

Lave, J. (1996). Teaching, as learning, in practice. *Mind, Culture, and Activity, 3*(3), 149–164.

Lave, J., & Wenger, E. (1991). *Situated learning: Legitimate peripheral participation*. Cambridge: Cambridge University Press.

Lee, H. C., & Jang, M. S. (2012). On the analysis and policy alternatives of an in-service teacher training on mathematics education. *Journal of the Korean School Mathematics, 15*(1), 171–182.

Lee, J. K., Lee, T. K., & Ha, M. (2013). Exploring the evolution patterns if trading zones appearing in the convergence of teachers' ideas: The case study of a learning community of teaching volunteers 'STEAM teacher community'. *Journal of the Korean Association for Research in Science Education, 33*(3), 1055–1086.

Putnam, R. T., & Borko, H. (2000). What do new views of knowledge and thinking have to say about research on teacher learning? *Educational Researcher, 29*(1), 4–15.

Schön, D. A. (1987). *Educating the reflective practitioner*. San Francisco: Jossey-Bass Publishers.

Seo, K. H. (2013). A community approach to teacher learning. *Journal of Educational Studies, 44* (3), 161–190.

Shulman, L. S., & Shulman, J. H. (2004). How and what teachers learn: A shifting perspective. *Journal of Curriculum Studies, 36*(2), 257–271.

Wenger, E. (1998). *Communities of practice: Learning, meaning, and identity*. Cambridge: Cambridge University Press.

Zazkis, R., & Liljedahl, P. (2008). *Teaching mathematics through storytelling*. Rotterdam: Sense Publishers.

Zeichner, K. M. (2001). *Educating reflective teachers for learner-centered education: possibilities and contradictions*. Plenary paper presented at the sixteenth Meeting of Encontro Nacional de Professores Universitários de Língua Inglesa, University of Londrina, Brazil.

Chapter 12
Classroom as a Site for Teacher Learning: Emergence of a Paradigm Shift in Mathematics Teacher Education in Pakistan

Anjum Halai

Abstract Pakistan is a country of about 180 million people with a big proportion being youth and young adults. The basic education system is therefore large with concomitant demands for an adequately qualified cadre of teachers. Historically the teacher education system in Pakistan has been weak, with a wide range of teacher preparation programmes and little if any availability of in-service teacher education. However, over the last two decades or so there has been a significant wave of reform in teacher preparation and in-service education in the country (e.g. HEC 2010). Situated within the context above, this paper reports on the case of an innovative field based in-service programme for mathematics teachers that recognized the significance of the classroom as a site for teacher learning. In this programme, the participating teachers tried new ideas into their classrooms and analysed the emerging issues with the mentor or the professional development tutor. Opportunity of engaging in reflection with the mentors created an 'in-between' space for the teachers to critique their practice and learn from it. The paper raises significant issues for policy and practice in teacher education.

Keywords Professional development · School-based teacher education · Third space

12.1 Introduction to the Context and Background

Pakistan came into being in August 1947 after independence from the British rule. It is a federation with four provinces, i.e. Punjab, Sindh, Balochistan and Khyber Pakhtunkhwa and four federally administered territories. The education system in

A. Halai (✉)
Institute for Educational Development, Aga Khan University, Karachi, Pakistan
e-mail: anjum.halai@aku.edu

B. Kaur et al. (eds.), *Professional Development of Mathematics Teachers*, Mathematics Education – An Asian Perspective,
DOI 10.1007/978-981-10-2598-3_12

the country is in the following stages: ECE[1] (3–5 years); primary education from (Class I–V, age 6–10 years); middle[2] school (Classes VI–VIII, age 11–14 years); high or secondary school (classes IX–X, age 15–16), higher secondary, (Classes XI–XII, age 17–18 years) and tertiary education. In 2010, as a result of the 18th Amendment to the constitution of Pakistan education was devolved to the provinces so that the provincial governments are free to devise education policy, planning and curriculum (GoS 2014). However, until the provinces develop their curriculum frameworks the National Curriculum 2006 (Ministry of Education 2006) is being followed throughout the country.

Mathematics is a compulsory subject up to the secondary school level. In the National Curriculum 2006, the mathematics content is organized mainly into five strands namely: number and number operations; measurement; geometry; data handling; and algebra. Within the effort to reform education in the country, a historical emphasis has been on improving the quality of science and mathematics teachers. For example, the National Education Policy (NEP) 2009 states, "In service teachers training in mathematics shall be given with due attention to developing conceptual understanding, procedural knowledge, problem solving and practical reasoning skills" (Ministry of Education 2009, p. 41). Likewise, several large-scale reform projects have focused specifically on capacity development of teachers of science and mathematics (Iqbal 2007).

Provision of teacher education both preservice and in-service is largely centred in the provinces with a wide variety of providers within the public sector and the private sector. The quality of teacher preparation and in-service education for mathematics and science teachers in the country has been an issue highlighted in a number of studies. For example, there is evidence that the quality of academic and professional preparation of male teachers is significantly better as compared to their female colleagues because the social and cultural constraints limit the opportunity for females to participate in higher education (Halai 2007a; Warwick and Reimers 1995); Teachers are not well versed in disciplinary knowledge in science and mathematics and this deficit was more pronounced in the case of rural teachers (e.g. Halai 2006; Iqbal 2007; SPDC 2003; Warwick and Reimers 1995); Teachers viewed mathematics as a fixed body of knowledge to be received from the experts (Halai 2007b; Amirali and Halai 2010).

More recently there have been extensive structural and policy level reforms in teacher education in the country undertaken by the government of Pakistan in collaboration with international partners. Among other elements these reforms include an orientation of professional development as opposed to training, introduction of a new 4-year B. Ed. degree programme with a strong component of practicum, a professional standards framework for teachers, review of the curriculum for teacher education so that it is aligned with the demands of the learners

---

[1]Historically Early Childhood Education (ECE) has not been a part of the formal education system. In some schools the traditional 'kutchi class' was offered to prepare learners for schooling.

[2]Elementary education includes primary and middle schooling.

in twenty-first century (Fazal et al. 2014; HEC 2010). The underlying position in these reforms is that preservice and in-service teacher education and teacher professional development are along a continuum of teacher learning that must continue over the life of a teacher's career and not be limited to the formal academic and professional qualification. Moreover, the strong emphasis on bringing teacher education closer to the site of practice is in line with the changes in the nature of teacher education more widely.

12.2 Teacher Learning in the Third Space

Over the last several decades, teacher education globally has seen a paradigm shift where the authority of the knowledge about teaching created in university settings was challenged in favour of a more hybrid source of knowledge about teaching that also valued the epistemology of the classroom (Zeichner 2010). As a consequence of this shift programmes of teacher preparation have formed collaboration between schools and universities or institutions of higher learning that offer teacher education. A consequence of these collaborative models of teacher preparation has been the creation of a third space for teacher learning as noted by Williams (2014), "Most teacher educators have previously worked as teachers in schools or other educational settings, and when they become teacher educators in universities, they are often involved in the supervision or mentoring of pre-service teachers in the field in the so-called 'third space'" (p. 315). Elsewhere it is noted that in initial teacher education the role of practicum takes on added significance and is meant to provide "in-between spaces for learning with reduced complexity as compared to the real world" (Wilson and I'Anson 2006, p. 360). However, for teacher learning in the 'in-between spaces' or the 'third space' to take place the practicum must be located within the framework of reflective practice for promoting growth in learning through analysis and questioning of issues emerging from practice (Schon 1987).

It is not just in teacher preparation programmes that the practicum or field-based component is seen as a crucial element for supporting growth in teachers' learning. In continuing teacher professional development and in-service teacher education also the role of practicum is significant but somewhat different. A critical difference is that in continuing teacher education teachers bring experience and knowledge from their teaching practice. They have already faced the challenges and complexities of the real world of schools and classrooms and bring valuable insights that could be built upon in the course of their continuing education. Having said that when practicing teachers are introduced to new ideas that require a shift in their practice they also face concerns and uncertainties akin to a novice and need support to manage the change (Anderson and Kumari 2009; Loucks-Horsley 1996). Additionally and perhaps more significantly, in traditional contexts where teaching is seen as knowledge transmission and the role of teacher as an expert who transfers knowledge, a paradigm shift is required to consider teaching and learning as a collaborative process, with the role of the teacher and learners modified accordingly

(Halai 2007b). Here again the role of practicum becomes significant for practicing teachers to problematize their practice through inquiry into teaching (Halai 2004, 2013).

Finally, there is an emerging body of work especially in mathematics education research, which maintains that teacher knowledge has multiple domains that need to be impacted in teacher education for the knowledge to have an effect on student achievement (Hill et al. 2005; Shulman 1987). Elsewhere, Ball et al. (2008) maintain, "Shulman's categories of content knowledge and pedagogical content knowledge can be subdivided into common content knowledge and specialized content knowledge, on the one hand, and knowledge of content and students and knowledge of content and teaching, on the other" (p. 391). These domains of knowledge recognize that there is specificity and difference in the way teachers hold knowledge of mathematics in teaching as compared to being able to do mathematics. The significance of this research for teacher education programmes is that the different domains of knowledge may be isolated in the seminars or in texts but in practice they are integrated (Kazima et al. 2008). By implication, the field component or practicum in teacher education enables the teachers to see these strands of knowledge as integrated.

12.3 The Case of the Advanced Diploma in Mathematics Education

The Aga Khan University Institute for Educational Development started its programmes in teacher education in 1994 initially in Pakistan and subsequently also in East Africa, with a 2-year master in education being its flagship programme and a range of other programmes including certificates, advanced diploma programmes and short needs-based trainings (Farah and Jaworski 2006). Typically, the graduates of the master's programme were mid-career teachers, after the masters degree they were known as the 'professional development teachers (PDTs)'. It was the PDTs who led the certificate, advanced diploma and other programmes with support and oversight from senior faculty (Shamim and Halai 2006). While details vary about the scope, depth of coverage and purpose of the various programmes and courses, they were underpinned by certain common principles and values about the nature of teacher education. Three key principles include as follows: (a) Teacher is a lifelong learner and agent of change; (b) Nature of knowledge is tentative and emergent; (c) Schools and classrooms are significant sites for teacher learning (Farah and Jaworski 2006; Khamis and Sammons 2004).

This chapter draws on the 1-year advanced diploma programme for mathematics teachers. The programme was structured in three phases. The first phase comprised of seminars held during summer break in the school year. Teachers were introduced to interactive approaches to teaching mathematics, use of concrete materials in mathematics and doing mathematics in small groups or pairs. Underlying purpose was to support teaching that was different from the usual transmission of knowledge and supported conceptual learning. Participants were introduced to strategies for reflection such as critical incident analysis and use of a reflective journal. They were also expected to undertake a small-scale action research project. The second phase comprised of a strong field component also known as the practicum. During this phase professional development tutors or mentors visited the teachers in the school to observe lessons, support post-lesson reflection and analysis and identify areas where teachers required further support. The third and the final phase comprised of synthesis and integration of learning, pulling together the action research reports and looking ahead to sustain the work initiated.

The data for this chapter comes from a systematic action research study by the author and other members of the team, teaching a cohort of the advanced diploma programme. The focus of the study was to understand the process of teacher development in the course of the advanced diploma programme (Halai 2004). The participants in the advanced diploma were 15 mathematics teachers from primary and lower secondary schools in Karachi. All 15 participants successfully completed the programme meeting the stringent requirement set by the university. As part of this action research study, the PDTs maintained detailed field notes of lessons observed, audio-recorded the conversation in post-observation analysis of lessons, collected all relevant documents including reflective journals maintained by the participating teachers lesson plans. The team of PDTs also maintained their own reflective journals and met regularly for a reflective dialogue among themselves. Provided below is the story of professional development through the illustrative case of Tehmina (pseudonym), a mathematics teacher with about 13 years of experience and a participant. The story traces the nuance and depth of development of knowledge and practice. It also raises significant questions for the policy and provision of teacher education in contexts of development.

12.4 Improving Mathematics Teaching: A Story of Professional Development

The opportunity of participation in an in-service teacher education programme as described above supported teachers in questioning their own practice and take steps to improve it. For example, Temina's engagement in the programme enabled her to pull back and look at her practice, and employ new approaches in her practice. She noted in her journal:

> In my 13 years I have not used this approach to teaching, which I am using now after my training at IED (Institute for Educational Development). In the sessions at IED I learnt while working on a simple activity on investigating area and perimeter that children do not learn just by telling or by making shapes on the blackboard. Children won't be able to learn until they do it themselves and until they get a chance to touch it themselves. (Tehmina's Journal entry, translated from Urdu)

She went on to plan lessons that incorporated use of concrete materials and interactive approaches in teaching and learning, and noted that these approaches were a change from her usual practice.

> I have been teaching in different schools for the past 13 years. In these 13 years I have observed that children depend solely on their teacher. Keeping this observation in mind certain questions come to my mind. If students don't understand, why is that so? Why do they feel shy in working in groups? Will using concrete materials influence their learning? If not then what factors would I need to keep in mind? (Tehmina's Journal entry, translated from Urdu)

However, the new approaches to teaching meant that unanticipated questions and issues came up in the course of the lesson. For example, when she employed concrete materials to teach mathematics a key issue was transition from concrete to abstract in learning mathematics.

Example 1 Lesson 1: This was a lesson in class VII on the topic of 'sets'. Tehmina gave out handfuls of tamarind seeds and date seeds to small groups of students and asked the students to make sets from the seeds, which she had given to them. Each group made two sets, one of tamarind seeds and the other of date seeds. She went around and monitored the work in progress. She next read from the textbook a formal definition of a set as follows: "A set is a collection of distinct objects" (Sheikh et al. 1998). Once the definition was given she wrote on the blackboard the following examples of sets and pointed out the convention of putting the brackets to signify a set and commas to separate the members of a set.

$A = \{a, b, c\} \qquad B = \{1, 2, 3, 4\}$

With reference to the set, A below the teacher instructed that the member 'a' was repeated and that it was not acceptable in Sets according to the definition.

$A = \{a, a, b, c\}$

Temina then asked the students, "how many members there were in each of the two sets?" (Referred to the sets students had made using the seeds), one student replied that it had "one member only". Tehmina accepted this response as correct without any further discussion.

Source: Halai (2004).

In the lesson extract in Example 1, above use of concrete materials in mathematics led to issues when learners had to transition from mathematics through concrete materials to the formal symbolic mathematics. The student's response 'one member only' was a naïve and counter intuitive observation about the property of sets. Presumably, the student's observation was arising from the teacher's instruction that the members of a set are not repeated. This observation by the students offered a 'teaching moment' to the teacher for additional discussion on what is meant by a 'distinct object' because the definition of a set as provided in the textbook noted 'A set is a collection of distinct objects'. The issue of distinctiveness is different in the case of concrete objects as compared to abstract or symbolic objects (e.g. alphabet, number). A discussion on the notion of 'distinct objects' as members of a set could potentially have opened up the space for an appropriate notation for a set with such concrete elements as its members.

A similar issue arose in the second lesson in Example 2.

Example 2 Lesson 2: This lesson was on 'equal sets and equivalent sets'. Tehmina explained to them that two sets with an equal number of elements were equivalent but were only equal if the elements were same and equal in number. Next, she gave handful of colourful buttons to small groups of students and asked them to make equal and equivalent sets with those buttons.

When the class worked with colourful buttons, Mehvish (a student) successfully formed a variety of sets and identified equivalent, equal and unequal sets. But, in the final, activity where students worked from their textbook to identify sets that were equal or equivalent, Mehvish identified the following two sets as equal:

$A = \{b,\ o,\ y\}$ $\qquad B = \{boy\}$

Tehmina used Mehvish's response as a discussion point in the class to illustrate the importance of the use of commas (,) in listing the members of a set.

Source: Halai (2004).

In the data extract above the two sets A and B were not equal because set A had three members while set B had one member. The commas were meant to separate each member. However, it appeared that Mehvish had not recognized the significance of commas in the formal notation. This could have been due to the fact that when making sets with concrete materials she did not need to put commas to separate the distinct members of the set. Hence, the very advantage in enabling students to learn through concrete materials as a stepping stone to the formal symbolism of mathematics could potentially become an issue if students are not supported to make the transition to the particular symbolic mathematical language. However, in this episode Tehmina did notice the students' incorrect response and acted upon it.

The episode also raised a broader issue about the use of concrete materials in teaching mathematics. Concrete materials like date seeds and tamarind seeds could potentially be an effective teaching resource if the mathematical intent inherent in the use of concrete material was not compromised. For example in the two episodes above, the transition from concrete to abstract was a moment where the teacher needed to take appropriate teaching decision. However, from the episode in Example 1 showed that the issue was not simply the transition from concrete to abstract. In the post-lesson conference Tehmina and the mentor worked through some mathematics tasks together and it emerged that her interpretation in the context of concrete materials that 'members of a set are not repeated' was the same way as the students interpreted it.

Reflecting on her teaching in the lesson on her Sets she wrote.

> During the lesson I had realised that those resource materials that I had used to teach sets were not appropriate for the purpose. Instead children were getting more confused with it. Hence, I had realised that I should have used some other teaching aids with clear and different things and which were more suitable for teaching sets. (Tehmina's Journal entry translated from Urdu).

To summarise, from the foregoing two episodes illustrated how the structures and processes of the post-lesson observation, entries in the reflective journal and working on mathematics tasks created opportunity for teacher growth and learning. These structures and processes between the PDT and the participant teacher were in the 'third or in-between' space, the PDT operated neither as a classroom teacher nor as a university academic but offered safe space for the teachers to analyse their practice and learn from it. Tehmina's story is illustrative of general patterns in the issues that arose and how the teachers worked with the PDTs as mentors.

12.5 Conclusions, Recommendations and Implications

From the data and analysis in the preceding section, it can be concluded that the practicum embedded in the in-service teacher education programme offered useful and productive opportunities to the teachers to convert their teaching practice into text, analyse it and learn from their practice. Toward this end, strategies for reflection were a necessary element of the practicum. Second, it can be concluded that the nature of issues arising were such that the pedagogic and the content knowledge were integrally entwined. Third, in contexts where teachers' academic preparation may not be strong the role of the mentor includes that of a subject expert.

However, the approach to teacher education raises some questions for policy and provision especially in the context of a large-scale education sector. To what extent is the approach of a mentor in an in-service three-phased programme applicable at scale? Hussain and Ali (2010) report on the "Cluster Based Mentoring Model" that was offered to more than 4000 teachers from public schools in Sindh and Balochistan. They maintain that a key feature of the success of the model was preparing a cadre of mentors with adequate knowledge, skills and attributes. They go on to explain that building a cadre of mentors required a careful selection and nurturing of mentors with supportive policies on deployment and transfer of teachers to ensure that capacity built on the ground was not dissipated (Hussain and Ali 2010, pp 70–80).

The data and analysis in this paper show qualitative and descriptive evidence of the impact of the approach to in-service teacher education as described above. Evidence of a sustained impact on the teachers' knowledge, skills and teaching practice of the field-based in-service education of the university is also noted in systematic impact evaluation of schools working with Aga Khan University (e.g.

Anderson and Kumari 2009; Khamis and Sammons 2004). Elsewhere, Quershi et al. (2011) studied through a quasi-experimental design with a sample size of 1109 students from classes VI–VIII, the impact on students' learning outcomes in science and mathematics on students who were in schools and classrooms of teachers who had undergone in-service education offered by the university. They maintain

> Results reveal that students in IED (Institute for Educational Development) intervention schools have scored higher in mathematics achievement tests across three grades as compared to their non-IED counterparts. This difference was significant for grades VI and VII ($p = 0.001$). Similar trends were noted with regard to science achievement tests for grades VI and VIII. However, the difference was found to be significant only for grade VIII ($p = 0.05$) (Quershi et al. 2011, p. 10).

To conclude, in-service teacher education and continuing professional development with a school-based practicum as its strong component offer a way forward for teacher education that is relevant to the reality of schools and classrooms. For teachers who may not have received adequate academic preparation such as in mathematics, the practicum also opened up opportunities for addressing issues related to content knowledge and pedagogic content knowledge. However, these programmes require a sustained and long-term investment in capacity building for a critical mass and a network that connects the schools and the university or institutions of higher learning.

References

Amirali, M., & Halai, A. (2010). Teachers' knowledge about the nature of mathematics: Survey of secondary school teacher in Pakistan. *Bulletin of Educational Research, 32*(2), 45–61.

Anderson, S., & Kumari, R. (2009). Continuous improvement in schools: Understanding the practice. *International Journal of Educational Development, 29*(3), 281–292.

Ball, D. L., Thames, M. H., & Phelps, G. (2008). Content knowledge for teaching: What makes it special? *Journal of Teacher Education, 59*, 389–407.

Farah, I., & Jaworski, B. (2006). *Partnerships in Educational Development. Oxford Studies in Comparative Education*. Oxford: Symposium Books.

Fazal, S., Khan, M. I., & Majoka, M. I., (2014). Teacher education in transition: A reform program in initial teacher education in Pakistan. In A. W. Wiseman & E. Anderson (Eds.) *Annual Review of Comparative and International Education 2014 (International Perspectives on Education and Society, Volume 25)* (pp. 357–378). Bingley: Emerald Group Publishing Limited.

GoS. (2014). Sindh education sector plan 2014–18. Karachi: Education and Literacy Department, Government of Sindh. Retrieved on December 17, 2015 from http://www.sindheducation.gov.pk/Contents/Menu/Final%20SESP.pdf

Halai, A. (2004). Action research to study impact: Is it possible? *International Journal Educational Action Research, 12*(6), 515–534.

Halai, A. (2006). Mentoring in-service teachers: Issues of role diversity. *Teaching and Teacher Education, 22*(6), 700–710.

Halai, A. (Ed.). (2007a). *Teacher status in Pakistan: A symposium*. Karachi: AKU-IED. ISBN 969-8073-13-8.

Halai, N. (2007b). Life history of a science teacher. In A. Halai (Ed.), *Teacher status in Pakistan: A symposium*. AKU-IED: Karachi. ISBN 969-8073-13-8.

Halai, A. (2013). Implementing curriculum change: Small steps towards a big change? In L. Tikly & A. Barrett (Eds.), *Implementing education quality in low income countries: Challenges for policy, practice and research* (pp. 168–180). London: Routledge.

HEC. (2010). Curriculum of education. B.Ed. 4-year degree programme (elementary, secondary), associate degree in education, MS/M.Ed. Education-revised 2010. Islamabad: Higher Education Commission. Retrieved from http://www.hec.gov.pk/InsideHEC/Divisions/AECA/CurriculumRevision/Documents/Education-2010.pdf

Hill, H. C., Rowan, B., & Ball, D. B. (2005). Effects of teachers' mathematical knowledge for teaching on student achievement. *American Educational Research Journal, 42*(2), 371–406.

Hussain, R., & Ali, S. (2010). Improving school teachers in Pakistan: Opportunities and challenges. *Improving Schools, 13*(1), 70–80. doi:10.1177/1365480209352404

Iqbal, H. M. (2007). Teachers' content knowledge and pedagogical competence: Issues for teacher status in Pakistan. In A. Halai (Ed.), *Teacher status in Pakistan: A symposium*. AKU-IED: Karachi. ISBN 969-8073-13-8.

Kazima, M., Pillay, V., & Adler, J. (2008). Mathematics for teaching: Observations from two case studies. *South African Journal of Education, 28*, 283–299.

Khamis, A., & Sammons, P. (2004). Development of a cadre of teacher educators: some lessons from Pakistan. *International Journal of Educational Development, 24*(3), 255–268.

Loucks-Horsley, S. (1996). The concerns-based adoption model (CBAM): A model for change in individuals. In National Standards & the Science Curriculum. Dubuque, Iowa: Kendall/Hunt Publishing Co.

Ministry of Education. (2006). National Curriculum for Mathematics Grades I–XII. Islamabad: Ministry of Education Government of Pakistan. http://www.ibe.unesco.org/curricula/pakistan/pk_al_mt_2006_eng.pdf

Ministry of Education. (2009). *National education policy 2009*. Islamabad: Ministry of Education Government of Pakistan.

Quershi, R., Bhutta, S., & Rodrigues, S. (2011). *Assessing impact of teachers'professional development on student learning outcomes in lower secondary schools in mathematics and science: A pilot impact study*. Report submitted to Aga Khan University Institute for Educational Development Pakistan.

Schon, D. (1987). *Educating the reflective practitioner: Toward a new design for teaching and learning in the professions*. San Francisco: Jossey-Bass.

Shamim, F., & Halai, A. (2006). Developing the professional development of teachers. Oxford Studies in Comparative Education (pp. 47–68), UK: Triangle.

Sheikh, M., Javed, M., & Dadlani, R. (1998). *Mathematics for class seven*. Jamshoro: Sind Textbook Board, Pakistan.

Shulman, L. S. (1987). Knowledge and teaching: Foundations of the new reforms. *Harvard Education Review, 57*, 1–27.

Social Policy Development Centre (SPDC) Survey (2003). Social development in Pakistan: The State of Education. Annual Review 2002–03. Karachi. SPDC.

Warwick, D.P., & Reimers, F., (1995). Gender and achievement. In: Hope or despair? Learning in Pakistan's Primary Schools. Praeger, London.

Williams, J. (2014). Teacher educator professional learning in the third space: Implications for identity and practice. *Journal of Teacher Education, 65*(4), 315–326. doi:10.1177/0022487114533128

Wilson, G., & I'Anson, J. (2006). Reframing the practicum: Constructing performative space in initial teacher education. *Teaching and Teacher Education, 22*, 353–361. doi:10.1016/j.tate.2005.11.006

Zeichner, K. (2010). Rethinking the connections between campus courses and field experiences in college- and university-based teacher education. *Journal of Teacher Education, 61*(1–2), 89–99.

Chapter 13
Teaching for Metacognition Project: Construction of Knowledge by Mathematics Teachers Working and Learning Collaboratively in Multitier Communities of Practice

Berinderjeet Kaur, Divya Bhardwaj and Lai Fong Wong

Abstract Teaching for metacognition project affirms a gradual shift in the centre of gravity away from the University-based, "supply side", "offline" forms of knowledge production conducted by university scholars for teachers towards an emergent school-based, demand-side, online, in situ forms of knowledge production conducted by teachers with support from fellow teachers, lead and senior teachers, and other experts such as university scholars and curriculum specialists. The project facilitates the participation of mathematics teachers in two-tier communities of practice. In this chapter, we describe the design of the project and the learning of two teams of teachers from two schools participating in the project. It is apparent from the findings that the teachers worked and learned collaboratively whilst participating in a first-tier and a second-tier community of practice. Their participation in the communities of practice enabled them to develop a deeper understanding of metacognition and also teaching for metacognition.

13.1 Introduction

Since 1998, in support of the Thinking Schools, Learning Nation vision (Goh 1997) of the Ministry of Education (MOE) in Singapore, all teachers are entitled to 100 h

B. Kaur (✉) · D. Bhardwaj
National Institute of Education, Nanyang Technological University, Singapore, Singapore
e-mail: berinderjeet.kaur@nie.edu.sg

D. Bhardwaj
e-mail: divya.bhardwaj@nie.edu.sg

L.F. Wong
Anderson Secondary School, Singapore, Singapore
e-mail: wong_lai_fong@moe.edu.sg

B. Kaur et al. (eds.), *Professional Development of Mathematics Teachers*, Mathematics Education – An Asian Perspective,
DOI 10.1007/978-981-10-2598-3_13

of training and core-upgrading courses each year. This entitlement is for teachers to keep abreast of current knowledge and skills. The professional development (PD) is funded by the MOE. There are many ways through which in-service mathematics teachers continue to develop themselves in Singapore. One of the ways is through participation in a research project facilitated by professors at the National Institute of Education, the sole institute for teacher education in Singapore. Two such past projects were Enhancing the Pedagogy of Mathematics Teachers (Teaching for Reasoning and Communication) (EPMT-RC) project (Kaur 2009, 2011) and the Think-Things-Through (T^3) project (Yeap and Ho 2009). In Singapore, these projects initiated the shift of PD activities from the "training model of PD" (Matos et al. 2009, p. 167) to the "hybrid model of PD" (Kaur 2011, p. 791).

In the training model of PD, teachers attend courses conducted by specialist officers from the mathematics Curriculum Planning and Development Division of the MOE or academics from the National Institute of Education. These courses are conducted for about 3 h per day spanning 4 to 10 consecutive days or days spread over some weeks. Almost always following the completion of such a course, there is no follow-up with the teachers about the use of the knowledge acquired and any impact that knowledge may have had on student achievement.

Research has shown that such courses are ineffective as teachers are likely to reject knowledge and skill requirements when (i) the requirements are imposed or encountered in the context of multiple, contradictory and overwhelming innovations; (ii) they are excluded from the development of the courses; (iii) PD is packaged in off-site courses or one-off workshops that are alien to the purposes and contexts of their work; or (iv) they experience them alone and are afraid of being criticized by colleagues or of being seen as elevating themselves on pedestals above them (Hargreaves 1995). Smylie (1989), in his survey of teachers' ratings of opportunities to learn in the US found that district-sponsored in-service workshops were at the bottom of the heap, ranked last out of 14 possibilities in terms of what teachers considered most valuable. Although such workshops are often accompanied by evaluations, seeking feedback on the duration, satisfaction, etc., efforts to measure what teachers learned have not been part of typical evaluation fare. In the same survey, Smylie found that teachers ranked direct classroom experience as their most important site for learning. Furthermore, for some teachers PD may not be an autonomous activity, i.e. chosen by a teacher in search of better ways of knowing and teaching mathematics (Castle and Aichele 1994).

The hybrid model of PD (Kaur 2011) integrates the "training model of PD" (Matos et al. 2009) with sustained support for teachers to integrate knowledge gained from the PD into their classroom practice. It exemplifies a critical development in the professional development of teachers in many parts of the world. This development reflects a gradual shift in the centre of gravity away from the University-based, "supply side", "offline" forms of knowledge production conducted by university researchers for teachers towards an emergent school-based, demand-side, online, in situ forms of knowledge production by teachers with support from university scholars.

In Singapore, the school mathematics curriculum is reviewed every 6–10 years. The last review was carried out in 2012. Since 1990, the framework for the school mathematics curriculum has been consistent. The primary goal of the school mathematics curriculum is mathematical problem-solving and the five interrelated components the framework places emphasis on; the learning of concepts, development of skills, attitudes, metacognition and mathematical processes such as thinking skills, heuristics, applications and modelling and reasoning, communication and connections (Ministry of Education 2012). In addition, the Ministry of Education (2010) identified competencies that have become increasingly important in the twenty-first century. These competencies include a confident person, a self-directed learner, an active contributor and a concerned citizen. In the US, the P21 Framework for twenty-first century learning has stressed that no twenty-first century skills implementation can be successful without developing core academic subject knowledge of mathematics and understanding among all students (Partnership for 21st Century Skills 2009).

Following the review of the curriculum in 2012, a group of teachers, university scholars and curriculum specialists examined the outcomes of three significant studies related to student achievement in mathematics. The studies are

(i) Programme for International Student Assessment (PISA) of 2009 (OECD 2010) and 2012 (OECD 2013);
(ii) Trends in International Mathematics and Science Study (TIMSS) of 2011 (Mullis et al. 2012; Kaur et al. 2013), and 2007 (Mullis et al. 2008; Kaur et al. 2012);
(iii) CORE 2 research at the National Institute of Education (NIE) by Hogan et al. (2013).

The findings of PISA and TIMSS showed that majority of Singapore students are very good in applying their knowledge in routine situations and this is definitely a consequence of what teachers do and use during their mathematics lessons. Hogan et al. (2013) found that there was a dominant use of performative tasks compared to knowledge-building tasks in grades 5 and 9 mathematics lessons that they studied. A performative task mainly entails the use of lower order thinking skills such as recall, comprehension and application of knowledge while a knowledge-building task calls for higher order thinking skills such as synthesis, evaluation and creation of knowledge. From the findings of the three studies, the group inferred that for students in Singapore to scale greater heights, teachers need to nurture metacognitive learners who are active and confident in constructing mathematical knowledge. Thus, a PD project—Teaching for Metacognition—was conceptualized as the greatest source of variance in the learning equation comes from teachers (Hattie 2009).

The project places emphasis on two key elements, knowledge-building tasks and teaching for metacognition, when used in tandem, create classroom discourse that facilitate students' active engagement in critical thinking, problem-solving, working collaboratively and articulating their thoughts and creating knowledge through their explorations. The goal of this chapter is to describe the Teaching for Metacognition project and examine the learning of two teams of teachers, from two schools participating in the project, in the two-tier communities of practice.

13.2 Teaching for Metacognition Project

Teaching for Metacognition project, is a hybrid model of PD (Kaur 2011). It is funded by the Academies Fund of the Ministry of Education and led by professors from the National Institute of Education, curriculum specialists from the Curriculum Planning and Development Division (Mathematics) of the Ministry of Education, a master teacher from the Academy of Singapore Teachers and a lead teacher from a secondary school. The aims of the project are threefold. The first is to provide teachers with knowledge and know-how of crafting knowledge-building tasks and how teachers may engage their students in metacognition during the learning of mathematics, i.e. metacognitive strategies. The second is to facilitate teaching for metacognition in the classrooms of teachers in the project. The third is to enthuse and support teachers to contribute towards the development of fellow mathematics teachers in Singapore and elsewhere.

13.2.1 Review of Literature

The conceptual framework of the project draws on research findings, specifically the characteristics of effective PD activities and teacher communities of practice. Relevant literature is reviewed in the following sections.

13.2.1.1 Successful PD Activities

High-quality and effective professional development programmes have been found to have a purpose as teachers are involved in shaping the foci of the programme so that it is related to their school work (Clarke 1994; Hawley and Valli 1999; Elmore 2002). These PD programmes are part of coherent programmes of teacher learning and development that support their instructional activities at school, such as adoption of new standards (Stiff 2002; Desimone 2009) and focus on how to teach and what to teach—the substance and the subject matter are keys (Stiff 2002; Desimone 2009). Lipowsky and Rzejak cited in Maaβ and Artigue (2013) noted that teachers viewed professional development initiatives as effective if they had clear relevance to their day-to-day teaching and the programmes had a clear focus on specific aspects of teaching or facilitation of student learning. Ball and Cohen (1999) have argued that "teachers' everyday work could become a source of constructive PD" (p. 6) through the development of a curriculum for professional learning that is grounded in the tasks, questions and problems of practice.

Such programmes include training, practice and feedback, and follow-up activities (Abdal-Haqq 1995). Ball (1996) claimed that the "most effective professional development model is thought to involve follow-up activities, usually in the form of long-term support, coaching in teachers' classrooms, or ongoing interactions with colleagues" (pp. 501–502). Effective PD programmes are sustained (Clarke 1994; Abdal-Haqq 1995; Hawley and Valli 1999; Elmore 2002; Stiff 2002; Borasi and Fonzi 2002; Desimone 2009) and embedded in teacher work (Clarke 1994; Abdal-Haqq 1995; Hawley and Valli 1999; Carpenter et al. 1999; Elmore 2002). Lipowsky and Rzejak noted that effective programmes are also intensive, combining learning-off-job in courses with learning-on-job in school. Teachers learn best when observing, being observed, planning for classroom implementation, reviewing student work, and presenting, leading and writing (Stiff 2002). Therefore, opportunities for teachers to engage in active learning are certainly related to effectiveness of PD (Wilson and Berne 1999; Desimone 2009). They also value the exchange of experiences with colleagues (Lipowsky and Rezak). In addition, collective participation by teachers from the same school, grade or department allow for a powerful form of teacher learning through prolonged interaction and discourse (Wilson and Berne 1999; Desimone 2009; Stiff 2002). PD programmes that foster collaboration have been found to be effective (Clarke 1994; Abdal-Haqq 1995; Hawley and Valli 1999; Elmore 2002; Borasi and Fonzi 2002).

13.2.1.2 Teacher Communities of Practice

Matos et al. (2009) noted that in teacher PD, learning should not be defined as the acquisition of knowledge of a propositional nature, but rather be conceptualized as being situated in forms of co-participation in the practices of teachers. Teachers participating in such learning may be said to belong to a community of practice (CoP) (Lave and Wenger 1991). Such a community does not exist when a group of teachers from several schools are interacting in a given setting such as attending a series of seminars or workshops, nor with groups of teachers in the same school who are teaching the same subject or year level but do not have mutual relationship and shared goals.

According to Wenger (1998), a CoP is a group of persons sharing the same practice. It has three key features: the members of a community of practice have a mutual enterprise; a shared commitment; and a common repertoire. This repertoire can contain material objects and stories that are shared by members of the community. By virtue of the design of a project like the EPMT-RC (Kaur 2011), with at least a group of teachers per school voluntarily participating in it with a shared set of goals and commitment to one another, within each school, the conditions were favourable for the development of a community of practice. Furthermore, as participants of the project, the teachers also had a shared sense of accountability towards the work of the community, in this case, the learning facilitated by the project. Contemporary sociocultural theory of learning acknowledges that learning involves increasing participation in a CoP composed of experts and novices (Lave and Wenger 1991). In the EPMT-RC project, while the teachers were participating in the project they were the novices and the university scholars were the experts;

subsequently as communities of practice at the school level enlarged, the teachers who had participated in the EPMT-RC project took on the role of experts and the newcomers were the novices.

Gueudet et al. (2013) caution that teachers working together on resources may sometimes not be a CoP. Rather they could be mere collectives. Gueudet et al. emphasize that advantageous conditions in terms of material provisions or time allocations are not enough to engage a dynamic towards a CoP. In addition, engaging such a dynamic requires teachers to work on common resources and to share professional knowledge and beliefs about the teaching of mathematics. In their study, they claim that turning collectives into communities requires the "development of a synergy between teachers and resources" (p. 1014) and that this may be facilitated by material conditions such as common meeting times for teachers to work on shared tasks with the support of external agents such as university scholars or experts in the field.

13.2.2 Design of the Project

The project has five significant features. The features are:

(i) *Content focus*
The project is focused on what to teach and how to teach (Stiff 2002; Desimone 2009). It was specific to the pedagogy of mathematics. This focus is similar to that of most in-service courses conducted for mathematics teachers in Singapore as the main objective of such courses is to introduce teachers to new initiatives that arise from curriculum revisions. The secondary school teachers participating in the project are working with mathematical content that was appropriate for the grade levels of their students.

(ii) *Coherence*
The project is coherent with the needs of the teachers. It focusses on teaching for metacognition which is one of the five components of the school mathematics framework that nurtures mathematical problem solvers. In addition, it also addresses a gap in instruction identified by Hogan et al. (2013), i.e. the disproportionate use of knowledge-building tasks by teachers to engage learners in higher order thinking during mathematics lessons. The project supports the instructional activities of teachers at school, such as the adoption of initiatives (Stiff 2002; Desimone 2009). Ball and Cohen (1999) have argued that classroom activities can form the basis of constructive professional development, and many other researchers have also determined that effective PD is embedded in teacher work (Clarke 1994; Abdal-Haqq 1995; Hawley and Valli 1999; Carpenter et al. 1999; Elmore 2002).

(iii) *Duration*
The duration of the project is 2 years and comprises three phases. Teachers attended training workshops for a semester, followed by a semester of school-based work guided and monitored by the university scholars (PD providers), followed by another year (2 semesters) of self-directed school-based work. The duration of the project is significantly longer than most in-service courses that mathematics teachers usually attend.

(iv) *Active learning*
The project engages teachers in active learning (Wilson and Berne 1999; Desimone 2009). It includes training, practice and feedback, and follow-up activities (Abdal-Haqq 1995), consistent with Stiff (2002), who suggested that teachers learn best when observing, planning for classroom implementation, reviewing student work, and presenting, leading and writing. As stated earlier, Ball (1996) also claimed that the most effective professional development model includes follow-up activities in the form of long-term support, coaching in teachers' classrooms, and ongoing interactions with colleagues.

(v) *Collective participation*
In the project there is collective participation at two levels—school and project. At the school level, at least four teachers, with pairs of teachers teaching the same grade year and mathematics programme are participating. These teachers work together during the training workshops and also at school when implementing their learning in their classrooms. At the project level, teachers also work together building their knowledge by participating in sessions during which they critiqued their peers' work, and shared their experiences and difficulties encountered during the implementation of their newly gained knowledge. Teachers in the project are participating in two-tier communities of practice, the school community of practice and the project community of practice.

13.2.3 Participants of the Project

Forty in-service secondary mathematics teachers from seven secondary schools in Singapore are participating in the project. The project is facilitated by a professor from the NIE, a research associate and a lead teacher from a secondary school.

13.2.4 Implementation of the Project

The project has three phases spread over 2 school years. A school year comprises two semesters, each of 20-week duration. Details of the phases are as follows:

13.2.4.1 Phase I

Duration of this phase is the first semester of the first year of the project (i.e. from January till May). The phase started with the participants completing the Pre-Intervention Teacher Survey. The survey sought from teachers their understanding about performative tasks, knowledge-building tasks, metacognition and teaching for metacognition. The findings of data from the survey were used to plan the knowledge-building workshops for the participants. Figure 13.1 shows the survey item on mathematical tasks and Table 13.1 the corresponding responses of the teachers.

From Table 13.1, it is apparent that teachers were using significantly more performative tasks compared to knowledge-building tasks during their lessons. This may have been a consequence of several factors, such as (i) the lack of knowledge-building tasks commonly found in textbooks used by the teachers; (ii) inability to craft knowledge-building tasks using textbook tasks that focus on direct application of knowledge; and lastly (iii) the push to develop procedural fluency after the introduction of concepts so as to perform routine tasks with ease during tests.

<table>
<tr><td colspan="4">Mathematical Tasks
The following are examples of performative and knowledge building tasks.</td></tr>
<tr><td colspan="4">Topic: Scales and Maps</td></tr>
<tr><td colspan="2">Performative task</td><td colspan="2">Knowledge-building task</td></tr>
<tr><td colspan="2">The scale of map A is 1: 40 000.
A rectangular field is 3 cm by 2 cm on the map. Find the actual area of the field in km^2.

If the area of the field is now represented on map B with scale 1: 20 000, what is the area on the map?</td><td colspan="2">The scale of map A is 1: 40 000.
A rectangular field is 3 cm by 2 cm on the map. Find the actual area of the field in km^2.

If the field is now represented on map B with scale 1: 20 000, without computing any area, explain how will the size of the field be different on map B.</td></tr>
<tr><td colspan="4">Topic: Quadratic graphs and graphical solutions of simple quadratic equations</td></tr>
<tr><td colspan="2">Performative task</td><td colspan="2">Knowledge-building task</td></tr>
<tr><td colspan="2">Draw the graph of $y = x^2 - 2x - 3$ for $-2 \leq x \leq 4$.
Hence solve the equation
$x^2 - 2x - 3 = -2$ graphically.</td><td colspan="2">Draw the graph of $y = x^2 - 2x - 3$ for $-2 \leq x \leq 4$.
Using your graph determine the number of solutions the equation $x^2 - 2x - 3 = a$ has.</td></tr>
<tr><td colspan="2">How often do you use performative tasks?
In ten consecutive lessons you would have used them</td><td colspan="2">How often do you use knowledge-building tasks? In ten consecutive lessons you would have used them</td></tr>
<tr><td>Please tick the appropriate response</td><td>√</td><td>Please tick the appropriate response</td><td>√</td></tr>
<tr><td>In all the lessons</td><td></td><td>In all the lessons</td><td></td></tr>
<tr><td>In 7 - 9 of the lessons</td><td></td><td>In 7 - 9 of the lessons</td><td></td></tr>
<tr><td>In 5 - 6 of the lessons</td><td></td><td>In 5 - 6 of the lessons</td><td></td></tr>
<tr><td>In 2 - 4 of the lessons</td><td></td><td>In 2 - 4 of the lessons</td><td></td></tr>
<tr><td>In 0 - 1 of the lessons</td><td></td><td>In 0 - 1 of the lessons</td><td></td></tr>
</table>

Fig. 13.1 Pre-intervention survey item on mathematical tasks used by teachers

Table 13.1 Responses of teachers to survey item on mathematical tasks

How often do you use **performative tasks**? In ten consecutive lessons, you would have used them		How often do you use **knowledge-building tasks**? In ten consecutive lessons, you would have used them	
N (%)		N (%)	
In all the lessons	20 (50.0)	In all the lessons	0 (0.0)
In 7–9 of the lessons	18 (45.0)	In 7–9 of the lessons	1 (2.5)
In 5–6 of the lessons	2 (5.0)	In 5–6 of the lessons	4 (10.0)
In 2–4 of the lessons	0 (0.0)	In 2–4 of the lessons	19 (47.5)
In 0–1 of the lessons	0 (0.0)	In 0–1 of the lessons	16 (40.0)

Our mathematics syllabus states that metacognition is "thinking about thinking".

a) What does metacognition mean to you? What is your understanding of metacognition?
b) Do you engage your students in metacognition during mathematics lessons? Yes / No.
If Yes, how do you engage your students in metacognition during mathematics lessons? Give an example of what you do.

Fig. 13.2 Pre-intervention survey item on teachers understanding of metacognition

Figure 13.2 shows the survey item that sought participants' perceptions about metacognition and teaching for metacognition.

Analysis of the qualitative data, of the survey item shown in Fig. 13.2, is reported in detail elsewhere (Kaur et al. 2016). From the analysis of the data it was apparent that teachers in the project had some knowledge about metacognition but their understanding was not comprehensive. Most of them generally tended to associate metacognition with higher order thinking and problem-solving. Only a few of them associated metacognition with awareness of thinking and reflection and critiquing of one's own thoughts. Thirty-two of the teachers claimed that they engaged their students in metacognition during mathematics lessons and all of them gave examples of how they did so. The examples showed that they did so by engaging their students in problem-solving and drawing on higher order thinking skills to solve mathematical tasks.

Seven 3-hour knowledge-building workshops were organized for the teachers. Table 13.2 shows the synopsis of the workshops.

13.2.4.2 Phase II

The second phase of the project was the second semester of the first year of the project. It was from July till November of the year. During this phase the school

Table 13.2 Synopsis of the workshops

Workshop	Synopsis of Workshop
1 and 2	*Performative and knowledge-building tasks* Teachers were introduced to performative and knowledge-building tasks by the professor and the lead teacher. They also demonstrated how a typical textbook task may be crafted into a knowledge-building one. Teachers were given four performative tasks and worked in school groups (according to their respective schools) to craft knowledge-building tasks. The first session drew to close with a show and tell activity during which all the groups shared their tasks with everyone in the project group and invited critique and suggestions During the second workshop, teachers in their school groups selected performative tasks from their textbooks and crafted knowledge-building tasks. Again they worked in their school groups, before sharing their knowledge-building tasks with all in the project group during the second half of the session. Following the two sessions, all the tasks created by the teachers in the project were made available to the project group through an e-portal. This was facilitated by the research associate of the project
3	*Teacher noticing* During this session, teachers watched a videorecorded lesson of the lead teacher in the project. In the lesson, the lead teacher was teaching for metacognition. During the first round of observing the lesson, the teachers were asked to note down on worksheet A (see Appendix A) what they observed After a short break the teachers were given worksheet B (see Appendix B) and asked to read it carefully before they viewed the videorecorded lesson once again. The prompts in worksheet B are adopted from McDuffie et al. (2014). During the second viewing of the videorecorded lesson, teachers noted down their observations on worksheet B. The session ended with a project group discussion on the four lenses that may be adopted for observing a lesson enacted by peers in the project
4–5	*Teaching for metacognition* In the first session, the paper Thinking about Thinking: Metacognition, Session 9 in The Learning Classroom: Theory into Practice—A Tele course for Teacher Education and Professional Development at the University of Stanford in the US (Darling-Hammond et al. 2001) was used as the resource for reading and discussion. Teachers worked in their school groups. Their deliberations were guided by the worksheet shown in Appendix C In the second session, teachers continued working in their school groups deliberating on aspects of metacognition. During the second half of the session, the school groups shared their examples of how teachers may engage students to reflect on what they know, direct their learning, create a culture of metacognition in the classroom and examples of strategies for developing metacognition. The session came to a close with teachers noting that there several strategies which may be used to engage students in metacognition [see list of strategies and examples in Kaur et al. (2016)]
6–7	*Planning a lesson to teach for metacognition* In the first session, teachers brainstormed as a project group and stated the mathematical norms and sociomathematical norms that would shape a knowledge and student-centred lesson. Next, they worked in their school groups and planned a lesson that they would teach in the next phase of the project. They were guided by the following instructions: • The lesson must use knowledge-building tasks and engage students in metacognitive strategies for learning • The lesson plan must state clearly (i) the lesson objectives (ii) the mathematical tasks used during the lesson (iii) the metacognitive strategies that would be developed (iv) the mathematical and sociocultural norms In the second session, the school groups engaged in a show and tell. They shared with the project group the specific instructional objective/s of their lesson, the mathematical tasks they planned to use, how they intended to engage the students in metacognition (stating specific strategies and key questions and prompts they planned to use), and what mathematical norms and sociomathematical norms guided their plan for the lesson. During the show and tell, the school groups invited critique and suggestions from fellow participants in the project group. Following the sharing, the school groups revised their lesson plans and prepared to teach the planned lesson during the next phase of the project

groups of the project worked collaboratively and implemented their planned lesson. They wrote a detailed lesson plan for the lesson they were carrying out. One teacher from the group taught the lesson to his/her students and the lesson was video-recorded. The school group met and viewed the lesson and prepared their presentation for the project group sharing meetings. Two project group meetings were held in October. During the project group meetings, the school group that presented solicited feedback from the project group. All participants in the project group except the teachers from the presenting school, participated in the feedback session. They used the four lens noticing feedback framework to give their feedback. The research team collected the feedback and the feedback was collated before it was sent back to the school group that presented.

Following the sharing sessions, the research team organized a meeting with every school group. Each meeting lasted between 2 and 3 h. A total of seven meetings were held. During the meetings the feedback from the project group was discussed and addressed. The feedback was very helpful as it provided the views of many more pairs of eyes reviewing the lesson. In addition, during the meetings the research team inducted the school group into a four-step approach to facilitate working and learning collaboratively when integrating their new knowledge into classroom practice. The four steps were as follows:

1. Plan and write a detailed plan of the lesson.
2. Enact and videorecord the lesson.
3. Watch the recorded lesson, compare it with the lesson plan and write the lesson narrative detailing the short comings and what the team would do differently the next time. A set of prompts were provided by the research team to guide the writing of the lesson narrative. The prompts were as follows:

   - Were the lesson objectives achieved? Was there any mismatch/deviation between the planned and enacted?
   - Were the mathematical tasks of knowledge-building type? How were the tasks enacted? Did they achieve the purpose they were intended for?
   - What were the metacognitive strategies that were developed? How were they developed? What challenges did the teacher encounter in developing them?
   - Did the teacher have any guiding mathematical norms that shaped the classroom discourse?
   - Did the teacher have any guiding sociomathematical norms that shaped the classroom interactions between the students, and also teacher–student?
   - What was the sequence of activities during the lessons? [e.g. teacher talk (demonstration), seat work, discussion/teacher talk (instructions), group work, student presentations, whole class discussion, etc.]
   - What was student engagement like during the lesson? [passive, active, problem-solving, explaining, problem posing, etc.]
   - Did the students say anything about the lesson? How similar or different it was from the teacher's normal lesson?
   - Would the teacher rate the lesson as one that taught for metacognition?

Table 13.3 Schedule of meetings to develop fellow teachers

School	Sharing session/conference
S1	Attended a conference in Korea in November [of the first year of the project]. Presented a paper showcasing their learning related to performative and knowledge-building tasks
S3	Organized a learning festival for staff in the school in November [of the first year of the project]. The teachers in the school group showcased lessons that taught for metacognition
S4	Teachers submitted a proposal to present their "Teaching for Metacognition Lessons" at the national Teachers Conference in June [of the second year of the project]

4. Write a reflection about the learning journey of the teacher's learning. Every member should do this individually, subsequently meet as a group and share with each other the reflections. The journal prompt was "*Describe in detail your learning journey during the planning, enacting and reviewing of your team's lesson that was carried out with the goal of teaching for metacognition*".

13.2.4.3 Phase III

Phase III of the project will be a year long and in the second year. Participants of the project will continue to work in their school groups and integrate their new knowledge into classroom practice. There will be periodic project meetings during which the school groups will showcase their lessons and invite critique and suggestions from the project group. In addition, the school groups will engage in activities through which they will contribute towards the development of fellow teachers both nationally and internationally. Table 13.3 shows a tentative schedule of activities that some schools have planned for developing fellow teachers who are not in the project, both nationally and internationally. At the time of writing this chapter the project had completed phases I and II only.

13.3 Learning of Teachers Working Collaboratively in Multitier Communities of Practice

In this section, we draw on the journals written by four teachers in the project to illustrate how the teachers in the project worked and learned collaboratively in the two-tier communities of practice. By virtue of the design of the project, with at least a group of teachers per school voluntarily participating in it with a shared set of goals and commitment to one another, within each school the conditions were favourable for the development of a community of practice. Furthermore, as participants of the project, the teachers also had a shared sense of accountability towards the work of the community, in this case, the learning facilitated by the project. In addition, there were school groups that formed the first-tier communities of practice and the project group that formed the second-tier community of practice. In the project there were seven first-tier communities of practice and only one second-tier community of practice. All the seven school groups together formed the second-tier community of practice while each school formed its own first-tier community of practice. The four teachers belong to two school groups and each group had enacted a lesson that taught for metacognition during phase II of the project. The lessons they enacted are described elsewhere (Kaur et al. 2015). The two teachers from the first school are S1T1 and S1T2 and the two teachers from the second school are S2T1 and S2T2. The teachers S1T1 and S1T2 were involved in the first lesson their school group enacted in their school while S2T1 and S2T2 did the same in their school.

13.3.1 Working and Learning in First-Tier Communities of Practice

During the knowledge-building workshops, teachers worked collaboratively in their school groups to craft knowledge-building tasks, clarify their understanding about metacognition and teaching for metacognition, and planning a lesson that teach for metacognition. From the following excerpts, it is apparent that the school group continued to work collaboratively to plan the lesson, enact it, view it and reflect on it during phase II of the project in their respective schools.

S1T1 *When the team decided on conducting a lesson using problem posing as our pedagogy, there was some resistance from the team members as we were unsure where and how to start. We discussed our challenges and used the guidelines given by the NIE professor to kickstart the project. The most challenging part was creating the rubrics for the students to assess the quality of the questions created by them. We managed to produce the rubrics by listing down the key components of the rubrics through our discussions and from other sources. … After the lesson was conducted, we reviewed the video clips taken in the classroom. We compared the lesson flow with our lesson plan. We realized that certain aspects in the lesson plan were not carried out or could have been carried out better. Aspects like missing keywords that were supposed to be said out, skipping a portion of the lesson plan due to time constraints,* etc. *The video evidence also showed us that there are improvements needed to be made for our future projects. Improvements like time allocation, group discussions and students' presentations. The video review also allowed us to check if we have met our project objectives and reflect on it.*

S1T2 *The sessions with the NIE professor and lead teacher prior to the planning of the actual lesson were instrumental in shaping what the group wanted to achieve in the lesson design. ... The team wanted to ensure that the lesson will be something innovative and one that the team has not designed before in the school. ... The team tried to also tie in the lesson design the department's pedagogy of Generative Activities, as the four design principles gel well with the concept of knowledge-building tasks and student-centred lesson. ... We had to do some reading up on problem-posing articles, and identified the types of activities that can be put into the lesson. At the same time, we were cognizant of the topics that are in the curriculum for the targeted class. ... The team prepared a lesson based on Sec 2 Algebra, and in the plan we highlighting good questions that were to be asked during the lesson.*

S2T1 *While planning, there were many considerations we had. ... these considerations posed many challenges. ... Bearing all the challenges in mind, we decided to carry out a knowledge-building task through linear graphs using ICT, focusing on the strategy "questioning by teacher" and "directed thinking".*

S2T2 *During the pre-planning stage, my team took into account the scheme of work when choosing the lesson idea for the project. We were also mindful about integrating other teaching methods, i.e. ICT, Learning Experience in order to enhance the collaboration and sharing amongst students. Therefore, we chose "Gradient of a Straight Line". When planning the lesson, we decided to focus on two metacognitive strategies, namely Awareness of knowledge and Monitoring activities during learning. We felt that these strategies could best support teaching and enhance metacognition in the students, and hence we developed our lesson around these strategies.*

13.3.2 Working and Learning in Second-Tier Community of Practice

During the knowledge-building workshops, teachers in the school groups presented their work to the project group and invited critique and suggestions. It was during these times that teachers worked and learned collaboratively in a second-tier community of practice. Furthermore, in phase II, when the school groups presented their lessons and invited critique and suggestions during the project group meetings, once again teachers worked and learned collaboratively in a second-tier community of practice. From the following excerpts, it is apparent that teachers valued the feedback they gathered during their participation in the second-tier community of practice as it contributed towards their deeper understanding of how to enact lessons that teach for metacognition.

S1T2 *During the last session of the workshops, the team was asked questions when they presented their proposal of problem posing. Inputs from the NIE professor, lead teacher and project group teachers provided a better understanding to what we wanted to achieve in our lesson. ... We learnt a lot when we reviewed the videorecorded lesson, and went through the other teachers' feedback [feedback from the project group that was collected during the school's group presentation in phase II]. For example, we realized that more opportunities can be given to students to play different roles, like presenting their feedback, besides being a critiquer. Also students could extend their learning by building their own 3D models.*

S2T1 *After reviewing, we realized that due to the nature of this lesson, "using discourse" seemed to be a better strategy to be used whereby students would need to first work and observe the lines drawn in different tasks before coming together in a pair to compare their answers. In the pair, each student is to explain how he/she obtains the observation and will have to justify and convince the other of the correct observation. This process helps students to concretize their thinking. ... Hence, amendments need to be made to the students' worksheet to add in a "food for thought" segment to allow discourse of how the values of* a *and* b *affect* $y = ax + b$.

13.4 Conclusion

The Teaching for Metacognition project, described in this chapter, illustrates a form of PD for mathematics teachers that is gaining momentum in Singapore. This is so as the PD is nestled in the classrooms of the teachers and addresses their needs. The three phases of the PD, namely: ***Learn*** (Acquisition and co-construction of knowledge), ***Apply*** (integrate new knowledge into classroom practice) and ***Teach*** (develop fellow teachers nationally and/or internationally) appear to make the engagement of teachers in PD holistic. In addition the working and learning of the teachers in tiers of communities of practice facilitates critique, suggestions and openness that builds synergy.

Appendix A

Teacher Noticing – Round 1

You have viewed the video record of Teacher A.

What did you notice?

List at least five observations as completely as possible.

<u>For example</u>: teacher talk – the teacher spoke in a monologue for extended periods of time.

My

observations ....

....

....

....

....

....

Appendix B

Teacher noticing—Round 2 You may use the following prompts to guide you in viewing the videorecord for the four lenses. The prompts are adopted from McDuffie et al. (2014)	
Teaching lens • **How does the teacher elicit students' thinking and respond**? – What opportunities does the teacher create for diverse learners to communicate their mathematical thinking—show what they know? – How does the teacher implement the task in a way that maintains or changes the cognitive demand?	*Learning Lens* • **What specific math understandings and/or confusions are indicated in students' work, talk, and/or behaviour**? – How do students communicate what their understandings and sense making of others' thinking? – In what ways does student engagement reflect conceptual and/or procedural learning?

(continued)

(continued)

– What resources and knowledge does the teacher use/draw upon to support students' math understanding?	– What resources or knowledge do students draw upon to understand and solve the math task?
Task lens • **What is the nature of the task/s used in the lesson**? – What makes this a good and/or problematic task? How could it be improved? What is/are the central math idea/s in this task? – How does the task make thinking visible? – What resources or knowledge does this task activate and/or connect to?	*Power and participation lens* • **Who participates**? • **Does the classroom culture value and encourage most students to speak, only a few, or only the teacher**? • **Where does the majority of the math "work" take place in the classroom**? – Who holds authority for knowing mathematics? Do some students hold more status than others? – What evidence indicates that differences in approaches and perspectives are/are not respected and valued?

Appendix C

Thinking about thinking: metacognition	
Metacognitive knowledge– Reflecting on What we know • Awareness of knowledge • Awareness of thinking • Awareness of thinking strategies *Write down examples of each for mathematics lessons*	Metacognitive regulation– Directing our learning • Planning approaches to tasks • Monitoring activities during learning • Checking outcomes *Write down examples of each for mathematics lessons*
A Culture of Metacognition in the Classroom *What conditions support a metacognitive classroom environment?*	Strategies for learning • Predicting outcomes/evaluating work • Questioning by the teacher/self-assessing • Self-questioning/selecting strategies • Using directed or selected thinking • Using discourse/critiquing/revising *Write down examples of each for mathematics lessons*

References

Abdal-Haqq, I. (1995). *Making time for teacher professional development (Digest 95-4)*. Washington, DC: ERIC Clearinghouse on Teaching and Teacher Education.

Ball, D. L. (1996). Teacher learning and the mathematics reforms: What do we think we know and what do we need to learn? *Phi Delta Kappan, 77*, 500–508.

Ball, D. L., & Cohen, D. K. (1999). Developing practice, developing practitioners: Towards a practice-based theory of professional education. In L. Darling-Hammond & G. Sykes (Eds.), *Teaching as the learning profession: Handbook of policy and practice* (pp. 3–32). San Francisco: Jossey-Bass.

Borasi, R., & Fonzi, J. (2002). Professional development that supports school mathematics reform. Foundations series of monographs for professionals in science, mathematics and technology education. Arlington, VA: National Science Foundation.

Carpenter, T. P., Fennema, E., Franke, M. L., Levi, L., & Empson, S. B. (1999). *Children's mathematics: Cognitively guided instruction*. Portsmouth, NH: Heinemann.

Castle, K., & Aichele, D. B. (1994). Professional development and teacher autonomy. In D. B. Aichele & A. F. Coxford (Eds.), *Professional development for teachers of mathematics* (pp. 1–8). Reston, VA: National Council of Teachers of Mathematics.

Clarke, D. (1994). Ten key principles from research for the professional development of mathematics teachers. In D. B. Aichele & A. F. Coxford (Eds.), *Professional development for teachers of mathematics* (pp. 37–48). Reston, VA: National Council of Teachers of Mathematics.

Darling-Hammond, L., Austin, K., Cheung, M., & Martin, D. (2001). Thinking about thinking: Metacognition. In L. D. Darling-Hammond, K. Austin, S. Orcutt, & J. Rosso (Eds.), *The learning classroom: Theory into practice* (pp. 156–172). A telecourse for teacher education and professional development. Stanford University school of Education.

Desimone, L. M. (2009). Improving impact studies on teachers' professional development: Toward better conceptualisations and measures. *Educational Researcher, 38*(3), 181–199.

Elmore, R. F. (2002). *Bridging the gap between standards and achievement: The imperative for professional development in education*. Washington, DC: Albert Shanker Institute.

Goh, C. T. (1997). Shaping our future: "Thinking Schools" and a "Learning Nation". *Speeches, 21* (3), 12–20. (Singapore: Ministry of Information and the Arts).

Gueudet, G., Pepin, B., & Trouche, L. (2013). Collective work with resources: An essential dimension for teacher documentation. *ZDM—The International Journal of Mathematics Education, 45*, 1003–1016.

Hargreaves, A. (1995). Development and desire: A postmodern perspective. In T. R. Guskey & M. Huberman (Eds.), *Professional development in education: New paradigms and practices* (pp. 9–34). New York: Teachers College Press.

Hattie, J. (2009). Visible learning. Routledge.

Hawley, W. D., & Valli, L. (1999). The essentials of effective professional development: A new consensus. In L. Darling-Hammond & G. Sykes (Eds.), *Teaching as the learning profession: Handbook of policy and practice* (pp. 127–150). San Francisco: Jossey-Bass.

Hogan, D., Towndrow, P., Chan, M., Kwek, D., & Rahim, R. A. (2013). *CRPP Core 2 research program: Core 2 interim final report*. Singapore: National Institute of Education.

Kaur, B. (2009). Enhancing the pedagogy of mathematics teachers (EPMT): An innovative professional development project for engaged learning. *The Mathematics Educator, 12*(1), 33–48.

Kaur, B. (2011). Enhancing the pedagogy of mathematics teachers (EPMT) project: A hybrid model of professional development. *ZDM—The International Journal on Mathematics Education, 43*(7), 791–803.

Kaur, B., Areepattamannil, S., & Boey, K. L. (2013). *Singapore's perspective: Highlights of TIMSS 2011*. Singapore: Centre for International Comparative Studies, National Institute of Education.

Kaur, B., Bhardwaj, D., & Wong, L.F. (2015). Developing Metacognitive skills of mathematics learners. In The Korean Society of Mathematical Education. (Ed.) *The Korean Society of Mathematics Education Proceedings of the 2015 International Conference on Mathematics Education* (pp. 237–245). Seoul, Korea: The Korean Society of Mathematical Education.

Kaur, B., Boey, K. L., Areepattamannil, S., & Chen, Q. (2012). *Singapore's perspective: Highlights of TIMSS 2007*. Singapore: Centre for International Comparative Studies, National Institute of Education.

Kaur, B., Wong, L.F., & Bhardwaj, D. (2016). Mathematics subject mastery—A must for developing 21st century skills. In P.C. Toh & B. Kaur (Eds.), *Developing 21st century competencies in the mathematics classroom*.World Scientific.

Lave, J., & Wenger, E. (1991). *Situated learning: Legitimate peripheral participation*. Cambridge: Cambridge University Press.

Maaβ, K., & Artique, M. (2013). Implementation of inquiry-based learning in day-to-day teaching: a synthesis. ZDM—The International Journal of. *Mathematics Education, 45*, 779–795.

Matos, J. F., Powell, A., & Sztajn, P. (2009). Mathematics teachers' professional development: Processes of learning in and from practice. In R. Even & D. L. Ball (Eds.), *The professional education and development of teachers of mathematics* (pp. 167–183). New York: Springer.

McDuffie, A. R., Foote, M. Q., Bolson, C., Turner, E. E., Aguirre, J. M., Bartell, T. G., et al. (2014). Using video analysis to support prospective K-8 teachers' noticing of students' multiple mathematical knowledge bases. *Journal of Mathematics Teacher Education, 17*, 245–270.

Ministry of Education, Singapore. (2010). MOE to enhance learning of 21st century competencies and strengthen art, music and physical education. Retrieved 5 Sept 2015 from www.moe.gov.sg

Ministry of Education, Singapore. (2012). *O-Level, N(A) Level, N(T) level mathematics teaching and learning syllabuses*. Singapore: Author.

Mullis, I. V. S., Martin, M. O., & Foy, P. (2008). *TIMSS 2007: International mathematics report*. Chestnut Hill, MA: TIMSS & PIRLS International Study Centre, Boston College.

Mullis, I. V. S., Martin, M. O., Foy, P., & Arora, A. (2012). *TIMSS 2011: International mathematics report*. Chestnut Hill, MA: TIMSS & PIRLS International Study Centre, Boston College.

OECD. (2010). *PISA 2009 results: What students know and can do: Student performance in reading, mathematics and science* (Vol. 1). OECD Publishing.

OECD. (2013). *PISA 2012 results: What students know and can do: Student performance in mathematics, reading and science* (Vol. 1). OECD Publishing.

Partnership for 21st Century Skills. (2009). *Assessment: A 21st century skills implementation guide*. Tucson, AZ: Author.

Smylie, M. A. (1989). Teachers' views of the effectiveness of sources of learning to teach. *Elementary School Journal, 89*, 543–558.

Stiff, L. V. (2002). March). Study shows high-quality professional development helps teachers most. *NCTM News Bulletin, 38*(7), 7.

Wenger, E. (1998). *Communities of practice: Learning, meaning and identity*. New York: Cambridge University Press.

Wilson, S. M., & Berne, J. (1999). Teacher learning and the acquisition of professional knowledge: An examination of research on contemporary professional development. *Review of Research in Education, 24*, 173–209.

Yeap, B. H., & Ho, S. Y. (2009). Teacher change in an informal professional development programme: The 4-I model. In K. Y. Wong, P. Y. Lee, Kaur, B., P. Y. Foong, & S. F. Ng (Eds.) *Mathematics education: The Singapore journey* (pp. 130–149). Singapore: World Scientific.

Chapter 14
Boundary Objects Within a Replacement Unit Strategy for Mathematics Teacher Development

Yew Hoong Leong, Eng Guan Tay, Tin Lam Toh, Romina Ann Soon Yap, Pee Choon Toh, Khiok Seng Quek and Jaguthsing Dindyal

Abstract We recognise that, for instructional innovations to take root in mathematics classrooms, curriculum redesign and teachers' professional development are two necessary and mutually-reinforcing processes: a redesigned curriculum needs to be seen as an improvement in order to facilitate teachers' buy-in—an ingredient for effective professional development; on the other hand, teachers' professional development content needs to be directed towards actual useable classroom implements through the enterprise of collaborative curriculum redesign. In this chapter, we examine the interaction between researchers and teachers in this collaborative enterprise through the metaphor of boundary crossing. In particular, we study a basic model of how "boundary objects" located within a "Replacement Unit" strategy interact to advance the goals of professional development.

Y.H. Leong (✉) · E.G. Tay · T.L. Toh · R.A.S. Yap · P.C. Toh · K.S. Quek · J. Dindyal
National Institute of Education, Nanyang Technological University, Singapore, Singapore
e-mail: yewhoong.leong@nie.edu.sg

E.G. Tay
e-mail: engguan.tay@nie.edu.sg

T.L. Toh
e-mail: tinlam.toh@nie.edu.sg

R.A.S. Yap
e-mail: rominayap@yahoo.com

P.C. Toh
e-mail: peechhon.tay@nie.edu.sg

K.S. Quek
e-mail: khiokseng.quek@nie.edu.sg

J. Dindyal
e-mail: jaguhthsing.dindyal@nie.edu.sg

B. Kaur et al. (eds.), *Professional Development of Mathematics Teachers*, Mathematics Education – An Asian Perspective,
DOI 10.1007/978-981-10-2598-3_14

14.1 Introduction

For the past five years, we have been working with various schools in Singapore towards the goal of making Mathematics Problem Solving (MPS) a more integral part of the mathematics curriculum of secondary students. As part of our efforts to directly impact mathematics classrooms, we design our teacher development activities to be sustained, collaborative, practice-based (i.e., concerned with issues related to practice), and close to the genuine instructional concerns of practice. We recognise that heightening the prominence of MPS in classrooms does not happen overnight, hence the need for sustained engagements. We also recognise that schools and teachers themselves should develop an ownership for the reform initiative for real change to take root, thus the importance of close collaborations with the teachers. Also, to impact classrooms directly, discussions with teachers should be about issues of direct concern to them. Thus, our professional development work is centred around issues of classroom practice.

In this kind of school-based teacher professional development activities, two groups of people with their respective cultural and professional traditions are brought together in collaboration: (1) Mathematics teacher educators who are usually mathematicians or mathematics educators working in the university; and (2) mathematics teachers who are based in schools. Metaphorically, the interactions between these two groups of professionals can be likened to a boundary crossing—a stepping over the perceived limits of one's professional 'world' into another social world. Seen in this way, the quality of teacher professional development can be studied through the activities around this boundary encounter.

14.2 Mathematics Teacher Professional Development as a Boundary Encounter

In defining professional development (PD) as a boundary encounter between researchers and teachers, we recognise that a natural boundary (Wenger 2000) exists between these two parties by virtue of the fact that they ordinarily operate in different domains and they possess different forms of expertise. Conventionally, researchers operate within universities or laboratories and are thought to possess more *formal knowledge* about practice; on the other hand, teachers operate within schools and possess more *practical knowledge* about practice (Cochran-Smith and Lytle 1999). PD provides a common domain or a boundary encounter where knowledge exchange, transfer, or creation can occur, which in theory can then lead to *boundary crossings* (Akkerman and Bakker 2011) of this knowledge benefitting both parties even as they return to operating in their usual domains.

While the potential for fruitful engagements within boundary encounters is evident, creating the conditions for successful boundary crossings is not straightforward. Three constructs highlighted by Sztajn et al. (2014) found within boundary encounters are helpful to consider when designing mathematics PD: Boundary brokers, boundary practices and boundary objects.

In teacher PD, the *boundary brokers* basically refer to the researchers and the teachers. As pointed out earlier, each of these two sets of individuals brings in different forms of expertise in the engagement. They also come in with a unique set of interests and expectations. These determine how each party acts as a broker for their respective domains.

Boundary practices, a result of boundary brokering, are the emerging or negotiated practices and norms which are carried out during the PD engagement. These define the manner in which knowledge is exchanged, transferred, or created.

Finally, *boundary objects* are the "representations of knowledge" (Sztajn et al. 2014) around which boundary practices are organised. Star and Griesemer (1989), from whom the term is originally attributed, describe boundary objects as

> objects which are both plastic enough to adapt to local needs and constraints of the several parties employing them, yet robust enough to maintain a common identity across sites … They may be abstract or concrete … The creation and management of boundary objects is key in developing and maintaining coherence across intersecting social worlds. (p. 393)

In the context of teacher PD, boundary objects may refer to the reform principles being adopted, or to more tangible classroom artefacts such as students' work. These objects do not necessarily hold the same meaning or value for the different parties in the boundary encounter. Cobb et al. (2009), for instance found that their use of students' work in PD "was, for the teachers, a tool for *retrospective assessment* of prior instruction rather than a resource for the *prospective planning* of future instruction" (pp. 187–188, emphasis in the original). The meaning and value of boundary objects, however, could be negotiated and transformed through appropriate boundary practices.

We find that the boundary metaphor provides us with an alternative image and understanding of teacher PD. It particularly draws attention to how the disposition of boundary brokers, the boundary objects involved in the boundary encounter (or the PD engagement), and the practices established for these encounters contribute to the success or failure of boundary crossings. However, the metaphor does not specify how these encounters should look like so that successful crossings can take place. Such a portrait can aid in designing and conducting PD.

In the kind of PD activities we engage with teachers for MPS over time, some of the constructs in the boundary metaphor have gained greater clarity to us. Based on our experience working with these schools in such PD settings, boundary brokers are researchers and teachers who have a commitment to teach MPS. Boundary practices refer to the structure and schedule of engagements for the purpose of professional learning that have become standard practice for both researchers and teachers. (These practices are elaborated in Sect. 14.6.) There is, however, a need for greater clarity about the boundary objects in such an enterprise and the way these objects interact to forward PD goals.

Thus, we conducted an inquiry focused on characterising the appropriate boundary objects for successful practice-based mathematics PD. In our study, we identified the boundary objects that we think will help us—researchers—interact productively with the teachers of the research school. We particularly asked the following questions: How do the boundary objects that we identified interact with each other? Are the interactions productive of teacher PD?

In this chapter, we would like to propose a possible theoretical framework addressing these questions. We then demonstrate the suitability of this framework based on our curriculum development work with one school.

14.3 Boundary Objects During Mathematics Teacher PD

In crafting a basic working model of the boundary objects during mathematics teacher PD, we drew on the previously highlighted differences in the expertise and intentions of the two main brokers—the researchers and the teachers—when they participate in the encounter and imagined the boundary objects that they would contribute to the engagement. It led us to view PD as involving *an interaction between two main boundary objects* in the boundary encounter, namely the *conceptualisations* of reform ideas and the *concretisations* of reform ideas for classroom use. Figure 14.1 illustrates how these two objects are conceived to interact in PD.

Conceptualisations are provisional models informed by theoretical ideas from the disciplinary domain or from pedagogical principles. They represent the knowledge about how mathematics reform is conceived to be applied in the classroom. Essentially, these conceptualisations often come from the researchers in the teacher PD initiative. They have their starting points from actual problems of practice. It is through these objects that researchers communicate to teachers what is thought of as a possible classroom scenario in light of the classroom reforms being promoted in PD.

On the other hand, concretisations of reform ideas are often more tangible. They embody the conceptualisations of the reform ideas as they are to be deployed for classroom use. They may come in the form of lesson plans, manipulatives or other classroom-ready instructional materials. Typically, the form which these concretisations take often come from the teachers as they are the ones who can decide what is appropriate for use in their classrooms. It is through these objects that teachers communicate to researchers how they interpret the reform being promoted for classroom implementation.

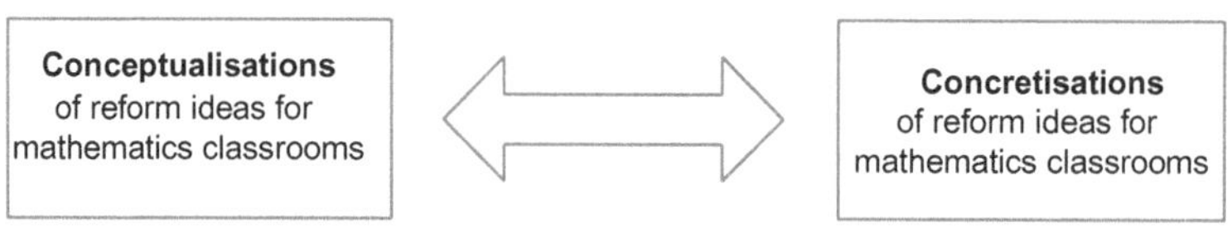

Fig. 14.1 Boundary objects in mathematics teacher PD

During PD, these two boundary objects 'interact' in the sense that revisions in one can affect changes in the other. They are thus objects that can undergo refinements throughout the PD process. We advance our proposed 2-object model by locating it within our school-based PD engagement with a secondary school in Singapore, which was focused on infusing MPS in the mathematics curriculum. We carried this out via the Replacement Unit strategy.

14.4 Situating the Boundary Objects Within the Replacement Unit (RU) Strategy

The Replacement Unit (RU) strategy is both a strategy for curricular redesign as well as a strategy for teacher PD. Its curricular redesign feature is exemplified by its main tenet of redesigning a whole unit or topic of study in the curriculum for the purpose of replacing the old one. In their work, Cohen and Hill (2000) also used the term "replacement unit". From a curricular redesign perspective, their definition of replacement units also coheres with our conception of it.

> Replacement units are curriculum modules designed to be consistent with reforms that center on specific topics… or sets of topics. [The units are devised] to be coherent and comprehensive in their exploration of mathematical topics – to truly replace an entire unit in mathematics texts, rather than just add in activities to existing curricula – and to be supportive of teacher as well as student learning. (p. 305)

Thus, in the RU strategy, the curricular content of a specified unit of study as well as the development of all the relevant classroom-ready instructional materials are redesigned and restructured according to a set of adopted reform principles. And in order for a replacement unit to be acceptable for teachers' implementation, the unit is redesigned without changing the time allocated for it in the school's teaching schedule.

To us, the RU strategy is also a teacher PD strategy. This PD goal is embedded in the overall RU strategy of teacher involvement in the redesign of the curriculum. An RU strategy engagement will typically begin with an identification of the unit of the curriculum to be replaced. An initial meeting is held between the design collaborators—both researchers and teachers—to discuss ideas. Usually, researchers lead in this meeting. Teachers will then redesign curriculum materials based on the agreed-upon ideas and these will undergo review and possible modifications in subsequent unit review meetings. Here, teachers present their designs for scrutiny. After the unit is implemented by the teachers, the RU strategy engagement concludes with an evaluation meeting to review the designed curriculum based on the classroom implementations. It is largely in this collaborative process of redesign that we view the RU strategy as a boundary encounter between researchers and teachers.

A RU spans 4–8 h of classroom time. We find this to be realistic and reasonable in terms of the scope of the work demanded for the redesign process and in terms of the required period of engagement for PD. Focusing on a larger curricular unit at once might be daunting simply for the amount of time and work it would demand on the design collaborators. On the other hand, a single lesson might not always capture the totality of the reform principles we want to promote because of its short duration. Focusing on a RU to trial (and re-trial, if necessary) and refine, thus, makes it a more manageable and meaningful engagement for curriculum development and PD.

In the described RU strategy, we identified the main boundary objects, which are representations of knowledge, as: (1) The ideas about the MPS-infused curriculum unit (Conceptualisations); and (2) the teachers' prepared instructional materials (Concretisations). In the rest of this chapter, we examine the interactions of these boundary objects within the context of designing a specific RU on probability. We begin by providing a brief background of the study.

14.5 Background of a Specific Case of RU-Design

The RU strategy engagement described herein was part of a broader design experiment entitled Mathematical Problem Solving for Everyone (MProSE). MProSE grew out of our desire to make MPS a more integral part of school mathematics in Singapore. This advocacy is consistent with the Singapore mathematics framework (Ministry of Education [MOE] 2006) which identified mathematical problem solving as "central to mathematics learning". Despite the existing framework, however, we found that actual implementation of MPS in Singapore schools was such that it only played a peripheral role in the students' mathematics learning rather than its envisioned central role (see for example Teong et al. 2009). That is, "problem solving" as enacted in mathematics classrooms was often relegated as a separate enrichment activity, sometimes packaged as an opportunity for students to encounter either recreational mathematics problems or problems that are considered as challenging. We also found that this situation was not unique to Singapore; in fact it was common for related international studies to report about how problem solving instruction still has not taken root in mathematics classrooms in spite of extensive past research (e.g. Doorman et al. 2007; Mok et al. 2005). This provided the motivation to embark on MProSE.

Currently in its fifth year of implementation, more details about MProSE, its implementation and outcomes can be obtained from our previous publications (Leong et al. 2014; Toh et al. 2011). In brief, MProSE can be described as having two main phases. The first phase, characterising our initial engagements with our partner secondary schools, was focused on the design and implementation of a 10-hour problem solving module which provided the students (and teachers) with the language and strategies used in problem solving. A central tool used in this module was the "Practical Worksheet" which was designed to habituate solvers to approach problem solving through the four stages of Pólya (1945/1973) model (i.e. Understand the Problem, Devise a Plan, Carry out the Plan and Check and Expand).

(A compressed version of the Practical Worksheet as adapted by the research school reported in this chapter is found in Appendix A.)

A key activity of the module was solving "MProSE problems". MProSE problems are problem solving tasks which are laid out in a Practical Worksheet. During class, students worked on these problems (individually or in small groups) for about 20 min, making use of the prompts in the Practical Worksheet to guide their MPS process. During this period, teachers were expected to monitor students' work and scaffold students' MPS processes.

Among the important outcomes of the first phase of MProSE was the common knowledge and understanding about MPS that was created among both teachers and students. It also resulted in a broader acceptance of a problem solving approach to teaching and learning mathematics. This paved the way for the next phase of MProSE which we are currently in the midst of.

The second phase of MProSE was focused on the actual infusion of problem solving in regular mathematics classes. This brought us further in making MPS a more integral part of the mathematics curriculum. Working within the regular curriculum, however, posed new challenges in our engagements with the schools. We adopted the RU strategy to maintain the manageability and meaningfulness of our engagements.

14.6 Implementing the RU Strategy in Eastpark Secondary School

Having previously established relationships with our research schools in the first phase of MProSE, Phase 2 formally started in Eastpark[1] Secondary School when the school expressed their willingness and readiness to proceed to infusion. They were then asked to identify a certain topic or unit of study which they found challenging to teach or which they wanted to infuse with problem solving. We describe in this section our implementation of the RU strategy in the school who expressed interest in transforming how they taught probability for their year eight classes. This unit spanned four one-hour lessons. It covered the definition of probability and computation of the probability of a single event.

The whole RU strategy—from curricular redesign, implementation and evaluation—spanned across 8 weeks. It started with an initial meeting[2] between the researchers and the relevant teachers of the school. In this meeting, the researchers presented their ideas about how the probability lesson infused with MPS can be

[1]Pseudonyms are used for the names of the school and school personnel cited in this paper.

[2]Collaborative meetings between the researchers and teachers were conducted in Eastpark Secondary School during the school's designated period for "PLC meetings"—a one-hour slot per week for teachers to discuss professional issues. Such a practice is becoming a norm in Singapore schools in line with the effort to develop professional learning communities (PLCs).

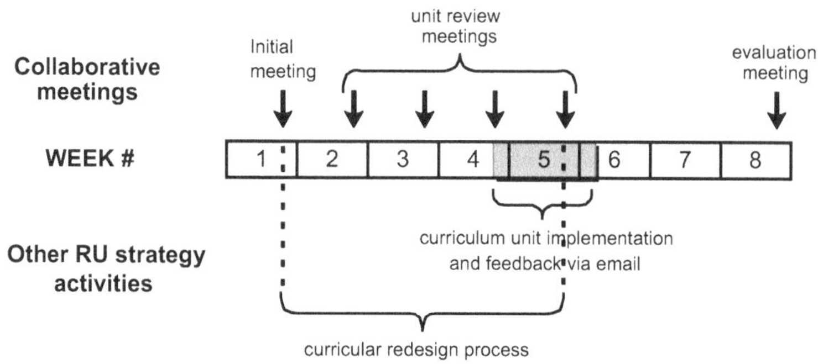

Fig. 14.2 The activities undertaken during the implementation of the RU strategy over the 8-week period

carried out over the prescribed four-lesson requirement. Teachers then independently designed the actual lessons, which in Eastpark, were embodied in instructional worksheets for the classroom. Three teachers were involved in the actual redesign—Cindy, Diane and Emma. Of the three, only Cindy and Diane were teaching the RU lessons. Emma was involved in her capacity as subject coordinator for the level.

In the weeks that followed, the researchers, the teachers and Jake, the head of the mathematics department, came together four times to review the curricular unit. During these meetings, teachers shared drafts of the worksheets which they designed, discussing their content and organisation and the rationale behind them. In the process, ideas that were previously presented by the researchers were revisited. Thus, this stage in the RU strategy was the period wherein both teachers and researchers were actively working together to produce workable and concrete curriculum materials that would meet the requirements and standard of both parties. The researchers would propose further modifications, but the final content and organisation of the worksheets were ultimately left for the teachers to decide.

In implementation, teachers were given autonomy to execute the lessons according to how they saw fit. We[3] made efforts to be kept posted of how the classes were implemented with minimal interrupts. We did not observe the teachers' implemented classes first hand, but the classes were video-recorded; we reviewed each lesson video before the next lesson of the RU. This enabled us to provide feedback to the teachers via email correspondences when we thought it was warranted.

The RU strategy process ended with an evaluation meeting of the whole process after the teachers have implemented all the lessons. Figure 14.2 provides a schematic of the different activities that took place over the said period of time.

---

[3]In the rest of this chapter, the plural personal pronoun is used to refer to the researchers, where applicable, for ease of reading.

14.7 Capturing the Interaction Between Boundary Objects in the RU Strategy

As identified in an earlier section, the specific boundary objects in this study are (1) the researchers' ideas about the MPS-infused probability unit, and (2) the teachers' prepared worksheets of the redesigned unit. The former, as mentioned, is a conceptualisation, while the latter is a concretisation.

From reviewing the activities involved in the implemented RU strategy, we sought to locate the interactions between the conceptualisation and concretisation occurring in the unit review meetings when the drafts of the worksheets designed were clarified alongside the conceived ideas for the MPS-infused lessons. In the subsequent sections of this chapter, we provide empirical evidence of this dynamic interaction between conceptualisation and concretisation.

We primarily reviewed the initial meeting through written documentations of what transpired to establish how the MPS-infused unit on probability was conceptualised. We identified the focus of the lessons, their intended organisation, and other specific items that were suggested when the lessons were conceptualised. We then reviewed the final worksheets that the teachers designed and used for their lessons and evaluated their fidelity to the initial conceptualisations according to the same attributes reviewed. We took note of the similarities and deviations from the original conceptualisation. Finally, we also reviewed the video recordings of the unit review meetings in order to have a better appreciation of the design process carried out to produce the worksheets.

14.8 The RU on the Topic of Probability

In this section, we present the findings of our analysis by first providing a picture of how the lessons were conceptualised and then describing how the worksheets concretised these conceptualisations. We then provide an overview of the nature of the discussions during the unit review meetings related to the design of the worksheets.

14.8.1 Conceptualisation of the RU Lessons

The topic of probability was chosen for infusion of units for these reasons: (1) the teachers requested for it as they found the topic hard to teach. A RU-design on this topic was thus a practical way to improve the teaching of the topic and provide the accompanying teacher PD to match the redesign; (2) due to the rich set of problems usually associated with probability, it is a topic amenable to an emphasis on MPS. In conceptualising the RU, we incorporated what we identified were three essentials that would facilitate students' understanding of probability better. First, the RU must be able to develop in students a motivation to quantify probability and not rely only on intuition. This involves instilling a discipline for careful study of the requirements when considering probability problems. We recognised that this should be developed progressively. Second, hand-in-hand with the first essential, the RU must be able to establish a common and formal language for communicating about probability in a progressive manner as well. This will facilitate the teaching and learning process. And finally, in line with the reform practice which we were advocating, the RU must feature infusion of problem solving to serve the purpose of deepening understanding in probability.

A description of how we conceived the four lessons is as follows:

Lesson One. The first lesson was essentially thought to provide students with an introduction to probability including some basic terms associated with it. We proposed beginning with tapping into the students' intuitive knowledge of probability as a lead-in. A series of questions about the likelihood of certain everyday situations occurring could be posed to students. At this point, informal language could be used. We suggested using a horizontal line, with the two opposite ends labelled as "unlikely" and "likely" respectively, which can be marked to indicate an estimation of the degree of likelihood of a certain event to happen.

It was then proposed to hold a short activity to illustrate how intuition can be faulty, thus providing a motivation for a more careful consideration when computing for probability. The specific activity suggested was the "Choose a number"[4] problem where students will be asked to choose and think of an integer between 1 and 200. [This is an adaptation to the "Birthday Problem".] The activity was meant to create a surprise at finding the chance of getting a match as being higher than expected. This was to provide the motivation for developing a more formal way of studying probability.

Lesson 1 was then proposed to proceed with introducing some standard language used in probability by defining "experiment", "outcome" and "event". And according to these introduced definitions, "probability of an event" can be defined provisionally as

$$\text{Probability of an event} = \frac{\text{no. of desired outcomes}}{\text{total no. of outcomes}} \tag{14.1}$$

Table 14.1 summarises how Lesson 1 was conceptualised according to the attributes identified in our analysis.

Lesson Two. We anticipated that students would have a tendency to still depend on intuition when responding to probability questions. Thus, in the second lesson we wanted students to develop a motivation for listing down all the outcomes in an experiment. We thought that this was a good opportunity for infusing MPS by allowing students to work on a MProSE problem featuring a probability situation

[4]Some of the problems as proposed during the pre-design meeting are given in Appendix B.

Table 14.1 Summary of how Lesson 1 was conceptualised

Lesson 1 attributes	Lesson 1 as conceptualised
Lesson focus	Introduction of probability and some basic terms
Lesson organisation	1. Start with everyday events to highlight the intuitive nature of probability 2. Illustrate with an activity how probability can be counter-intuitive 3. Proceed with introducing the basic standard terms in probability and define the probability using these terms (See Definition 1) 4. Conclude with exercises that will not only provide students with practice on finding probabilities but also on making use of the standard language of outcomes
Specific suggestions	• Use of a horizontal line to indicate the degree of likelihood of a certain event to occur as part of tapping into the students' intuitive understanding of probability • Use of the "Choose a number" question for establishing how probability can be counter-intuitive (referred to in #2 above)

where the intuitive outcomes were not equally likely. We thus provided some options that teachers could consider using—the "three children" problem, the "passenger seat" problem, and the "loaded die" problem. According to our conceptualisation, students would be given some time to independently work on this problem. After which a synthesis of the solution of the problem can be done, followed by introducing more formal probability language, in particular using "sample space" to denote the collection of all possible outcomes, and adjusting the definition of "probability" as found below to reflect this progression.

$$\text{Probability of an event} = \frac{\text{no. of favourable outcomes}}{\text{no. of outcomes in the sample space}} \tag{14.2}$$

The importance of listing down the sample space with equally likely outcomes is to be emphasised, making reference to the students' work with the problem posed at the start of the lesson. Table 14.2 summarises how Lesson 2 was conceptualised.

Lesson Three. In the third lesson, the definition of probability was to be extended to account for the case where the number of outcomes of a sample space was infinite. We thought that this was an important concept to include to provide the unit a better sense of completeness from a mathematical discipline perspective. To begin the lesson, it was suggested that teachers can pose an appropriate probability problem that can illustrate the inadequacy of the previously established definition of probability. Specific suggestions for the problem to pose included a spinner problem and a dart problem where the corresponding board has unequal sectors. After leading students to an intuitive solution to the problem, an extended definition of probability can then be formally introduced. In particular,

Table 14.2 Summary of how Lesson 2 was conceptualised

Lesson 2 attributes	Lesson 2 as conceptualised
Lesson focus	Introduce the concept of outcomes being equally likely and the formal term "sample space"
Lesson organisation	1. Start with an MProSE problem on probability which features an experiment with intuitive outcomes which are not all equally likely, thus motivating students to further re-examine their intuitive approach to solving probability problems 2. Introduce new formal language which then defines probability according to certain outcomes in the sample space (See Definition 2) 3. Proceed with exercises that will help develop a discipline for identifying outcomes in the sample space and the event prior to determining the probability
Specific suggestions	• Three possible MProSE problems were suggested—the "three children" problem, the "passenger seat" problem and the "loaded die" problem

$$\text{Probability of an event} = \frac{\text{measure of favourable event}}{\text{measure of sample space}} \tag{14.3}$$

Table 14.3 summarises how Lesson 3 was conceptualised.

Lesson Four. For the last lesson, we could include a MProSE problem that would utilise the concepts of probability taught in the previous lessons. We shared the "phoney Russian roulette" problem as a possible MProSE problem that could be used where students could be given time to independently work on the task before the discussion of the solution. It was then suggested that the lesson could end by returning to the "choose a number problem" introduced in Lesson 1. A summary of the conceptualisation of Lesson 4 is found in Table 14.4.

14.8.2 Lessons as Concretised in the RU Worksheets

The final worksheets used for the RU lessons were found to keep close fidelity to how they were conceptualised. We did, however, notice some deviations or additions. In the description that follows, we present an overview of these similarities and deviations.

Lesson One. The final worksheet prepared for Lesson 1 followed closely to how it was initially conceptualised. All the attributes as proposed in the conceptualisation were adapted (see Table 14.1). Figure 14.3 shows the actual portion of the worksheet for the first portion of the lesson.

Table 14.3 Summary of how Lesson 3 was conceptualised

Lesson 3 attributes	Lesson 3 as conceptualised
Lesson focus	Extension of the probability concept to situations that involve non-discrete measure
Lesson organisation	1. Start with posing a problem that involves non-discrete measure as a motivation to extend the probability definition 2. Introduce the extended definition of probability involving finding "measure" instead of "number" (See Definition 3) 3. Proceed with a mixed set of exercises that reinforce earlier taught concepts
Specific suggestions	• Using a "spinner" or "dart" problem where the board has unequal sectors

Table 14.4 Summary of how Lesson 4 was conceptualised

Lesson 4 attributes	Lesson 4 as conceptualised
Lesson focus	Consolidation of students' knowledge about probability
Lesson organisation	1. Start with a suitable MProSE problem 2. Close the unit by recalling and solving a simpler version of the problem presented in Lesson 1
Specific suggestions	• Use the "Phoney Russian roulette" problem for the MProSE problem • If the "Choose a number" problem was adopted in Lesson 1, a simpler problem can be solved instead

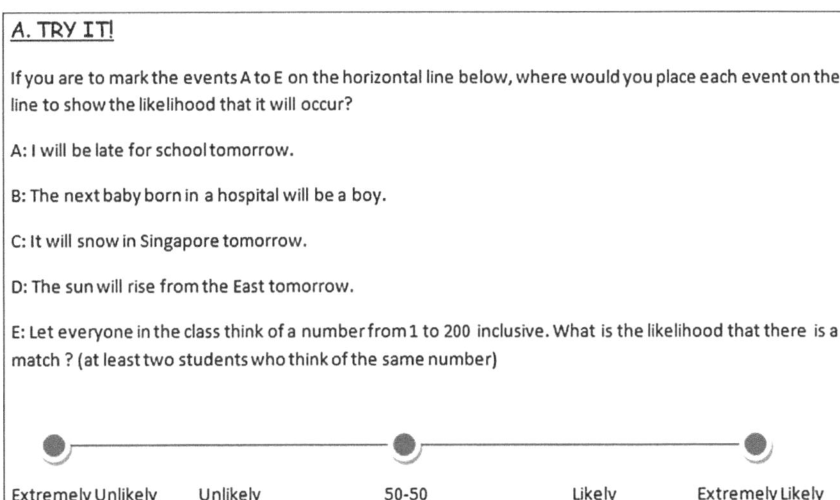
A. TRY IT!

If you are to mark the events A to E on the horizontal line below, where would you place each event on the line to show the likelihood that it will occur?

A: I will be late for school tomorrow.

B: The next baby born in a hospital will be a boy.

C: It will snow in Singapore tomorrow.

D: The sun will rise from the East tomorrow.

E: Let everyone in the class think of a number from 1 to 200 inclusive. What is the likelihood that there is a match ? (at least two students who think of the same number)

Extremely Unlikely Unlikely 50-50 Likely Extremely Likely

Fig. 14.3 Part A of the final worksheet designed in Lesson 1

C. CLASS PRACTICE

1. A fair die is thrown. Find the probability of getting an odd number.

Experiment	Act: Observe:
Outcomes	
Event	

Related Practice: CASCO
Pg 257 Q 4,5

Fig. 14.4 The first item in the exercises provided in the final worksheet designed for Lesson 1

The definition of the probability of an event was presented as follows:

$$\text{Probability of an event} = \frac{\text{no. of favourable outcomes}}{\text{total no. of possible outcomes}} \quad (14.4)$$

Finally, the worksheet concluded with a set of exercises where the first few questions did not only require students to find the probability of a certain event, but also require them to identify and characterise the experiment, the possible outcomes and the outcomes of the indicated event. The first of these problems is shown in Fig. 14.4.

Items that were in the concretisation that were not specified in the conceptualisation were these:

An exercise in the worksheet shown in Fig. 14.5 addressed P(not E) = 1 − P(E); also, the teachers included a worksheet for homework consisting of four questions that were meant to reinforce the concepts covered in Lesson 1.

Lesson Two. Again, all the attributes in Table 14.2 were adopted, with these minor modifications: (1) The first worksheet was a modified "three children" problem printed on the Practical Worksheet; (2) a homework sheet was also prepared for this lesson reinforcing the lessons learned; and (3) the second worksheet provided a recall of the definitions previously discussed but also brought in further use of formal notation and language by introducing the term "sample space", thus redefining the terms "event" and "probability" in relation to the sample space. Thus, the definition of the probability of an event was now presented as

$$\mathrm{P(E)} = \frac{\mathrm{n(E)}}{\mathrm{n(S)}} \quad (14.5)$$

Lesson Three. The worksheet for Lesson 3 began with a recall of the definition of probability of an even as presented in the previous lesson (i.e., P(E) = n(E)/n(S)).

5. A bag contains 3 blue balls, 4 white balls and 1 red ball. A ball is drawn at random from the bag.

Find the probability that the ball drawn is

(a) blue

(b) pink

(c) a ball

(d) not blue

Related Practice: CASCO
Pg 261 Q12, 18

$0 \le P(E) \le 1$ for any event E

Impossible Event P(E) =

Certain Event P(E) =

$\therefore$ P(not E) =

*6. There are 15 red fish and n white fish in an aquarium tank. If a fish is selected at random, the probability of selecting a white fish is $\frac{4}{9}$. Find the value of n.

Fig. 14.5 The first item in the exercises provided in the final worksheet designed for Lesson 1

It then proceeded with two problems wherein the outcomes required a different measure. (The first problem involved area, while the second problem involved lengths.) It was followed by an extended definition of the probability of an event, where "measure" was used instead of "number", exactly as that previously set in Definition 3. A set of practice questions then followed, covering different types of situations.

Lesson Four. The worksheet prepared for the last lesson consisted only of the "Phoney Russian roulette" problem which was printed on a Practical Worksheet. Thus, closing the unit by recalling and solving the "Choose a number" from Lesson 1 was not carried out as suggested in the conceptualisation.

14.8.3 Discussions During the Collaboration Sessions

The final worksheets adhered closely to how the unit was initially conceptualised. This can be taken as an indicator of how the two boundary objects interacted with each other. In this case, the worksheets drew heavily on the ideas for an MPS-infused unit on probability.

It may be argued, however, that the close correspondence of the worksheets to the conceptualisation can be interpreted differently. For instance, it can be said that teachers were just blindly or robotically complying with the ideas that they were presented with. Only through reviewing the original drafts of the worksheets and the unit review meetings can one fully appreciate the amount of clarification, negotiation, and transformation that went into the production of the final worksheets—and hence the dynamic interaction with the conceptualisation of the RU.

In reviewing the meetings, we took note of the various discussions that were undertaken in relation to the design of the worksheet. We identified four, not necessarily disjoint, aspects of the worksheet which were modified as a result of these discussions—sequence, coverage, task and wording.

Sequence. The first drafts of the worksheets did not all reflect the sequence of parts as found in the final worksheets. For example in Lesson 1, the "Choose a number" problem was initially designed to be a MProSE problem to be administered at the very beginning of the class instead of it being a short question or activity following the everyday probability questions as seen in Fig. 14.3. In Lesson 2, the "Three children" problem was thought to come after formally introducing the concepts of sample space and equal likelihood of outcomes. Discussions that led to changes to sequence of parts of the worksheet drew on consideration of the lesson rationale and the time constraints among other things.

Coverage. The content covered in the worksheets also saw some changes. A major change of course was the extension of the probability concept covered in Lesson 3. A more specific example was how the initial draft of the Lesson 3 worksheets included a statement about how the probability of a point lying on the boundary of certain shaded region is zero. It was agreed, however, that while this was something that teachers need to know, it might be prudent not to introduce this concept to the students unless they bring it up so as to not confuse them.

Task. The MProSE problems, exercises and examples appearing in the worksheets were carefully reviewed in the collaboration meetings. Their efficacy in producing the desired effect and their appropriateness were considered. Teachers would explain the inclusion of a certain task according to how they thought it can introduce a concept. (e.g. "The fifth [example] is to lead them on to impossible events and certain events".) or reinforce a discipline they wanted students to develop (e.g. "I just thought that this one… will make them… list down all the possibilities".).

The collective review of tasks would sometimes lead to revisions in or certain decisions about the task. For example, the ends of the horizontal line in Fig. 14.3 were originally labelled as "most likely and "most unlikely". But it was pointed out that using the word "most" seemed to imply a need to compare events. Thus, "most" was replaced with "extremely". Another example is how the original textbox in the exercise shown in Fig. 14.4 did not require the specification of the components of the experiment (i.e. the "Act" and what to "Observe"). But when it was recalled that it was something they wanted to reinforce, this was added. The decision to not include a simpler version of the "Choose a number" problem was also a result of a collective review of the conceptualised lesson. After trying out different versions of a simpler problem, revisiting the lesson goals, and considering the time allotted for the lesson, it was decided that this portion of the conceptualised lesson may not be necessary.

Wording. Apart from wording of the tasks, wording of the different statements and definitions appearing in the worksheets were also carefully scrutinised. Ensuring accuracy was a major consideration. For example, the beginning of the statement which defined "event" was corrected from "The event is" to "An event is". Another example was the attention placed on ensuring that the probability definition included a clause about equal likelihood of outcomes when it was appropriate. Consistency was also a concern. This was reflected when efforts were made to make sure that definitions appearing across worksheets were identical whenever it was called for.

14.9 Discussion and Conclusion

In this study, drawing upon the boundary crossing metaphor, we sought to examine our proposed 2-object model which consists of the objects of conceptualisation and concretisation (Fig. 14.1). Each of the objects draws upon the forte of the respective "worlds" of researchers and teachers. Our interest lies in the potential of interaction, especially the conceptualisation → concretisation direction, between the objects.

As it turned out, the concretisations retain a high level of fidelity to the initial conceptualisation, with substantial teacher inputs that reflect their active sense-making for productive adoption in their classrooms. We interpret this to mean that the boundary encounter as facilitated by the two boundary objects provides a feasible platform for the teachers' PD in the teaching of this topic.

As a thought experiment, we may suppose an alternative model where the PD was designed as mainly a one-object encounter. Specifically, suppose that the PD engagement was designed in such a way that the conceptualisations and the concretisations were packaged as a single object designed mainly by the researchers meant for the teachers' ready use in their classrooms. In fact, this approach basically summarises the model of most reform engagements. In traditional reform models, the full works of curriculum development—which includes conceptualisation, design and production of curricular and instructional materials—are undertaken by "experts" engaged directly by policy-makers. In this model, teachers play at most a bit-part role of contributing feedback to the near-final documents. Mostly, teachers' role is seen mainly as implementation of the designed curriculum. This results in teachers not having a direct stake in the concretisation which in itself is a form of personalised reconceptualization—which we think is necessary capacity-building to carry out the reform.

On the other hand, another conception of a one-object model is one where teachers take on the primary role of packaging both the conceptualisation and concretisation. This is essentially the model of the now-famous Lesson Study enterprise. In Lesson Study, teachers undertake the primary role of leading in conceptualisation and concretisation while university researchers (known as Knowledgeable Others) are invited in their teams. Because of their 'guest' status, their roles are normally limited to post-design advice that involves minor refinements to the teachers' plans. In this case, however, the researchers' higher vantage point in disciplinary and pedagogical matters is not sufficiently harnessed for significant improvements in curriculum design and teacher PD.

Admittedly, this study has not explored the full power of the boundary crossing metaphor. We have limited our field of study to the interaction of only the two objects of (mainly researcher-led) conceptualisation and (mainly teacher-owned) concretisation. Another important boundary object that can potentially advance communication between the two worlds is Actualisation—(video) images of how the concretised materials are harnessed during classroom enactment. We can expect this object to present a different set of affordances and tensions in the PD engagements. This is an example of how research building upon the boundary metaphor can proceed along. It is an area that we would like to explore in future research.

Appendix A: A compressed version of the Practical Worksheet Used in Eastpark Secondary School

Instructions

1. You may proceed to complete the worksheet doing stages I–IV.
2. If you wish, you have 15 min to solve the problem without explicitly using Polya's model. Do your work in the space for Stage III.
   - If you are stuck after 15 min, use Polya's model and complete all the stages I–IV.
   - If you can solve the problem, you must proceed to do stage IV—Check and Expand.
   - You may have to return to this section a few times. Number each attempt to understand the problem accordingly as Attempt 1, Attempt 2, etc.

PROBLEM OF THE DAY:

Stage I: Understand the Problem

(a) Use some heuristics such as Draw a Diagram, Restate the problem, Use Suitable Numbers, etc., to help you.
(b) I have understood the problem. (Circle your agreement below)

Strongly Disagree		Neutral	Strongly Agree	
1	2	3	4	5

Stage II & III: Devise a Plan and Carry it out

(a) State your plan clearly, for example: (i) Use suitable Numbers and Look for Patterns; or (ii) Find the areas of all smaller triangles and work out their ratios.
(b) Number each plan as Plan 1, Plan 2, etc.
(c) Carry out the plan that you have stated.
(d) Write down in the *Control* column, the key points where you make a decision or observation, for e.g. go back to check, try something else, look for resources or totally abandon the plan.

Stage IV: Check and Expand

(a) Write down how you checked your solution.
(b) Write down a sketch of any alternative solution(s) that you can think of.
(c) Give one or two adaptations, extensions or generalisations of the problem. Explain succinctly whether your solution structure will work on them.

Appendix B

Some of the specific problems suggested during the RU strategy

Problem name	Problem details
Choose a number	Write down an integer between 1 and 200. What is the probability/chance that there is a match?
Three children	Given a family with three children, what is the probability that the family has three boys?
Passenger seat	Mr. and Mrs. Tay and their son John goes into their family car. Mr. and Mrs. Tay can drive but John cannot. What is the probability of Mr. Tay sitting in the passenger seat? (Apart from the driver seat that must be filled, the other passengers can choose to sit in any of the remaining seats in the car.)

(continued)

(continued)

Problem name	Problem details
Loaded die	In an unbiased die, three faces are painted "1", two faces are painted "2", and the last face is painted "3". Find the probability that when the die is rolled, "2" is obtained
Phoney Russian roulette	Two bullets are placed in two consecutive chambers of a 6-chamber revolver. The cylinder is then spun. Two persons play a safe version of Russian Roulette. The first points a gun at his hand phone and pulls the trigger. The shot is blank. Suppose you are the second person and it is now your turn to point the gun at your hand phone and pull the trigger. Should you pull the trigger or spin the cylinder another time before pulling the trigger?

References

Akkerman, S. F., & Bakker, A. (2011). Boundary crossing and boundary objects. *Review of Educational Research, 81*(2), 132–169. doi:10.3102/0034654311404435.

Cobb, P., Zhao, Q., & Dean, C. (2009). Conducting design experiments to support teachers' learning: A reflection from the field. *Journal of the Learning Sciences, 18*, 165–199.

Cochran-Smith, M., & Lytle, S. L. (1999). Relationships of knowledge and practice: Teacher learning in communities. *Review of Research in Education, 24*, 249–305.

Cohen, D. K., & Hill, H. C. (2000). Instructional policy and classroom performance: The mathematics reform in California. *Teachers College Record, 102*(2), 294–343.

Doorman, M., Drijvers, P., Dekker, T., van den Heuvel-Panhuizen, M., de Lange, J., & Wijers, M. (2007). Problem solving as a challenge for mathematics education in the Netherlands. *ZDM—The International Journal on Mathematics Education, 39*, 405–418.

Leong, Y. H., Tay, E. G., Quek, K. S., Toh, T. L., Toh, P. C., Dindyal, J., et al. (2014). *Making mathematics more practical: Implementation in the schools*. Singapore: World Scientific.

Ministry of Education [MOE]. (2006). *Secondary mathematics syllabuses*. Singapore.

Mok, I. A. C., Cai, J., & Fung, A. T. F. (2005). *Teaching mathematics through problem solving: Struggles a Hong Kong teacher faces*. Paper presented at the The 3rd East Asia Regional Conference on Mathematics Education, Shanghai, China.

Pólya, G. (1945/1973). *How to solve it: A new aspect of mathematical method* (2nd ed.). New Jersey: Princeton University Press.

Star, S. L., & Griesemer, J. R. (1989). Institutional ecology, 'translations' and boundary objects: Amateurs and professionals in Berkeley's Museum of Vertebrate Zoology, 1907–39. *Social Studies of Science, 19*(3), 387–420. doi:10.1177/030631289019003001.

Sztajn, P., Wilson, P. H., Edgington, C., Myers, M., & Teachers, P. (2014). Mathematics professional development as design for boundary encounters. *ZDM—The International Journal on Mathematics Education, 46*, 201–212.

Teong, S. K., Hedberg, J. G., Ho, K. F., Lioe, L. T., Tiong, J. Y. S., Wong, K. Y., Fang, Y., et al. (2009). Developing the repertoire of heuristics for mathematical problem solving, project 1: Establishing baseline data for mathematical problem solving practices in Singapore schools. Final technical report for project CRP 1/04 JH. Singapore: Centre for Research in Pedagogy and Practice.

Toh, T. L., Quek, K. S., Leong, Y. H., Dindyal, J., & Tay, E. G. (2011). *Making mathematics practical: An approach to problem solving*. Singapore: World Scientific.

Wenger, E. (2000). Communities of practice and social learning systems. *Organization, 7*(2), 225–246. doi:10.1177/135050840072002.

Chapter 15
Facilitating Professional Growth of Taiwanese In-service Mathematics Teachers Through an Innovative School-Based Program

Fou-Lai Lin, Hui-Yu Hsu and Jian-Cheng Chen

Abstract The chapter begins with the present challenges that mathematics educators in Taiwan are facing. It outlines the hierarchical structure of in-service teacher development and highlights how the top-down approach often adopted may not address the needs of classroom teachers. The authors of the chapter next describe an innovative school-based program, Lighten-up School-Based Program (LUSBP), they initiated for facilitating professional growth of mathematics teachers in Taiwan. The core for LUSBP is that all tiers of educators, teachers, and students learn through active participation whilst interacting with each other. The project employs a design-based approach with teachers as designers who learn from the process of creating tasks, enacting tasks with classroom students, and revising tasks based on students' learning. The school-based model enables the creation of a friendly learning environment where teachers take it for granted to make changes and are willing to share their experiences with one another. The outcomes of LUSBP are positive and hold promise for the future.

Keywords Professional growth · In-service mathematics teachers · School-based professional development program · Design-based professional development program

---

F.-L. Lin (✉)
National Taiwan Normal University, Taipei City, Taiwan
e-mail: linfl@ntnu.edu.tw

H.-Y. Hsu
National Hsinchu University of Education, Hsinchu City, Taiwan
e-mail: huiyuhsu@mail.nhcue.edu.tw

J.-C. Chen
Ming Chi University of Technology, New Taipei City, Taiwan
e-mail: janchen@mail.mcut.edu.tw

B. Kaur et al. (eds.), *Professional Development of Mathematics Teachers*, Mathematics Education – An Asian Perspective,
DOI 10.1007/978-981-10-2598-3_15

15.1 Main Challenge in Taiwan Mathematics Education

Students' performance on large-scale mathematics assessments in East Asian countries can be generally categorized into two contrasting groups (Mullis et al. 2008; Mullis et al. 2012; OECD 2005). One group comprising countries such as Taiwan, Hong Kong, Korea, and Singapore has consistently ranked on top on those assessments. However, students in this group usually have low confidence in and keep negative attitude towards the learning of mathematics. The other contrasting group comprising countries such as Thailand, Indonesia, Philippines, and Malaysia has students with high confidence and positive attitude but their students performed much lower than those in the first group. While mathematics education aims to nurture students in both mathematics competence and positive learning attitude, each group of countries has to face their specific educational challenges.

The challenges in Taiwan involving the prevailing phenomenon of students' low confidence in and negative learning attitudes towards mathematics are strongly related to the examination culture. Lin and Tsao (1999) pointed out that examination-driven instruction often asks students to practice as many examples as possible and does not undertake any explorations or extensions because of the time consideration. The drill and routine practices lead to memorization and decrease students' interest in learning mathematics. Lin (2009) further indicated that Taiwanese teachers often think that teacher-centered instruction (talk and chalk) is necessary for successful learning, such an instructional situation results in the lack of opportunities for students to think mathematics actively. Examination-driven environment is also very likely to influence novice teachers in shaping their teaching styles at the beginning of their careers in schools.

We believe that curiosity and the ability to stand for a proposition are the key to the succession of mathematics learning. It is important for students to learn how to generate examples in dual relation with mathematics concepts and mathematics results (Michener 1978). In this regard, Taiwan Central Counseling Team (CCT) proposed a nation-wide professional development project, namely Lighten-Up School-Based Program (LUSBP), using a design-based approach with the aim to facilitate in-service mathematics teachers to improve their teaching and promote students' active thinking and learning power. Here, learning power refers to the supple and nimble minds so that students will be able to learn whatever they need to (Claxton 2002).

15.2 Central Counseling Team (CCT) in Taiwan

Taiwanese official teacher counseling organization mainly involves two hierarchical systems: Central Counseling Team (CCT) and Local Counseling Teams (LCT). CCT is affiliated to Ministry of Education, whereas LCT is affiliated to local administrative divisions (e.g., municipality and county). For CCT, Ministry of Education (MoE) (2008) clearly stated the purposes of establishing the system are

(1) to assist the propaganda and implementation of policies related to curriculum and instruction; (2) to promote professional growth of teachers and their knowledge for teaching; (3) to enhance the quality of classroom teaching and student learning; and (4) to construct a curriculum development model based on the integration of research and practice as well as to build up a tertiary instruction counseling system.

To achieve those purposes, the approach that CCT from 2002 to 2011 usually employed was to hold professional development workshops with the topics that were either new ideas learned from literature (e.g., collaborative learning) or those that have the potential to overcome the challenges specific to Taiwan education (e.g., examination-driven instruction). CCT invited experienced educators, usually experienced mathematics teacher educator-researchers (MTE-Rs), to implement the workshops and required CCT teacher members or those selected from LCT to participate in. The participating CCT and LCT teachers trained then become seeded teachers who can in turn be the educators or mentor teachers to help local school teachers learn the new ideas or policies. In this regard, CCT plays an intermediate role of coordinating and connecting between MoE and LCT, and supports LCT to facilitate school teachers to adopt the new policies and implement new instructional ideas in their classrooms.

The intermediate role of CCT for coordination and connection is important. However, the tertiary instruction counseling system from MoE, CCT, LCT, then to school teachers very likely creates educative challenges. The main challenge has to do with the tertiary instruction counseling system which is more inclined to a top-down model where CCT followed policies and then determined the topics that school teachers need to learn. Then, the sequential transformation from MoE, CCT, LCT, and then to school teachers makes it possible that school teachers are passive receivers, who inactively receive the knowledge and policies that MoE and CCT attempt to convey. Recognizing that the transfer of knowledge from educators, teachers, to students in professional development is not a linear and one-way process in which the solutions to problems encountered in teaching and learning can be directly obtained. This transferring process is complex, cyclic, and sophisticated; involving interplays with multiple tiers of participants (Lesh 2003; Lesh et al. 2007; Lesh and Kelly 2000). Thus, it is highly possible that the top-down approach leads school teachers think that they can know the solutions from educators instead of formulating instructional strategies specific to their pedagogical challenges themselves. A notable example for the top-down approach in Taiwan is the implementation of Constructivism Teaching Approach around 2000, in that case classroom teachers only knew the general ideas of the new teaching approach from the one-shot workshops arranged by MoE. However, teachers were lack of competence of transforming the learned ideas into actual classroom teaching. Another challenge may be due to the way of assigning one teacher from each individual school to participate in professional development workshops. Even the assigned teachers are willing to change their classroom teaching based on the new ideas learned from the workshops, they may not be able to convince other school teachers to do so. Also the challenge can be related to the knowledgeability and identity of LCT in facilitating school teachers to solve a diversity of pedagogical challenges encountered. All of those challenges can downplay the effectiveness and efficiency of professional development workshops and result in limited professional changes of in-service teachers.

Additionally, the evaluation for professional development workshops often used quantitative reports. For example, how many workshops have been held each semester? How many times do LCT members consult with school teachers? Fundamental issues regarding teacher growth, the improvement of classroom teaching, and students' learning quality may not be revealed by this evaluation approach accordingly. In order to solve the challenges mentioned previously, we (the authors of this chapter) proposed an innovative project for facilitating professional growth of in-service mathematics teachers. The project is described in the following sections.

15.3 Innovative Professional Development Program: Lighten-up School-Based Program (LUSBP)

The conceptualization of the LUSBP is shaped by three main aspects, namely research experiences, school-based model and active participation of schools, and co-operation of university faculties in Taiwan. In the following sections we elaborate on each of them.

15.3.1 Research Experiences

15.3.1.1 Experiences with the Implementation of Design-Based Professional Development Workshops for Experienced Mathematics Teachers

The main foundation that supports us to set up the nation-wide professional development project is our experiences with the implementation of design-based professional development workshops for experienced mathematics teachers for three years. The aim for the workshops is to enhance active learning: not only for students to learn mathematics actively, but also for teachers and the educators to make changes for enhancing their pedagogical power and educative power respectively. Of importance is to create a learning environment where students can experience the essence of mathematics learning. The essence of mathematics learning is rooted in the well-organized body of mathematics knowledge, involving enculturation of students' minds through environmental activities (e.g., explaining) (Bishop 1991), perception of how mathematics is developed and formulated (Lakatos 1976), and the understanding of the origins of mathematics from an epistemological perspective (Freudenthal 1983). The core idea is to enable students to observe and derive crucial mathematics ideas and meaning from reality, and develop the mathematics sense to solve problems in- and out-of-mathematics.

The rationale for the design-based professional development is the design research paradigm (Cobb et al. 2003). To implement design approach, we employed a three-layer structure including grand theory, intermediate framework, and design tool (Gravemeijer 1994; Ruthven et al. 2009). The latter two serve to coordinate and contextualize the theoretical insights from grand theories about the epistemological and cognitive perspectives into the designs and the study of their operations. In order to facilitate teachers to design tasks, we also propose three entries including misconceptions, propositions in school mathematics, and mathematical facts; by which teachers can much easier initiate a plan for the design. Five researchers including the authors played roles of designers and critical commentators for arranging workshops that better facilitate professional growth of experienced teachers through a task-design approach. The researchers chose conjecturing and diagnostic teaching as the themes for the workshops as they, on the one hand, can easier motivate students to think mathematics actively, and, on the other hand, can detect students' learning problems and then design tasks that can solve the problems accordingly.

Being task designers, teachers in the professional development workshops have opportunities to intensely explore curriculum materials and students' learning as well as to incorporate professional development materials into their designs; all of which become important sources for improving their knowledge for teaching. To help teachers design tasks, relevant theories and strategies with corresponding examples were elaborated. During the workshops, teachers were required to initiate tasks, present their designs in professional development for obtaining feedback from educators and peer teachers, enact the designs with students, and revise the work accordingly. The whole process offered participating teachers opportunities to detect and challenge their individual pedagogical problems. The professional development programs were led by the first author, an experienced educator who mastered in both research in mathematics education and teaching practices in classrooms. Because of his expertise in both research and teaching, the educator was able to elaborate the research and theories in association with student cognition and provide directions for the refinement of tasks.

The experiences with implementation of the professional development workshops for three years enable us to propose three keys that have great impact on the succession of teachers' growth in profession.

(i) Getting started: Formulating pedagogical problems by teachers

Cooney (1994) reported that teachers may not see any conflicts in their teaching and call for the attention to how teachers formulate and then solve their pedagogical problems for the growth in profession. Cooney stated the notion of pedagogical power, indicating that it involves recognizing conditions and constraints of classroom situations, assessing the consequences of possible actions, and then determining what actions are the best for the particular situations in classroom teaching. As change of teachers must be "motivated by some need to adopt or engage or reject" (Reid and Zack 2010, p. 372), teachers' extrinsic and intrinsic motivation that triggers them to get started in formulating pedagogical problems is of importance. After formulating pedagogical problems, the follow-up stage is how teachers come up with plan to challenge their pedagogical problems.

(ii) Coordination as mechanism for teacher growth in profession

Reflection and enactment are currently the key trust in teacher education (Jaworski 1993; Smyth 1989). We propose coordination mechanism and argue that it can clearly identify the quality of teachers' reflection and enactment in terms of how they construct novelty for their professional growth. Particularly, we concern what sources of information that teachers perceive and reflect in professional communities, and how they coordinate the multiple sources and then instigate the follow-up actions accordingly. We also identify the crucial components involved in coordination mechanism. The elaboration of the mechanism with examples will be published soon by Lin, Hsu, Yang, and Chen.

(iii) Elaborating theoretical ideas at proper moments and in proper ways

We recognize the importance of theoretical ideas for teachers to structure their teaching and to identify students' learning difficulties. However, theories are not usually appreciated by school teachers as they may directly respond that theories or principles are far away from their teaching practices and argued that they only want something that can be directly used in their teaching without any further work. Our research experiences show that elaborating theoretical ideas for teachers necessities educators' power for communicating with educative phenomenology, reasoning onto emergent models and theories about educative phenomenology, and maintaining a dialectical connection between practice and research (Yang et al. 2015). Through supporting and challenging, educators facilitate teachers to understand the meaning of these theoretical ideas and their contributions to classroom teaching at proper moments and in proper ways.

15.3.1.2 Experiences with an Integrated Research Project

Another important foundation for LUSBP is a three-year design-based research project founded by Taiwan national science council for professional development with seven sub-projects in seven universities. After implementing design-based professional development workshops for three years, the first author launched the research project and called for cooperation in Taiwan. As a result, seven university faculties who master in professional development and have different research expertise participated in the project. Carrying out the integrated research project gives us confidence in initiating and implementing LUSBP. One main reason is that those faculties participating in the project are also the MTE-Rs for LUSBP. Another is that the execution of the project offers us the opportunity to comprehensively reflect the nuances that are keys to the succession of a nation-wide program.

15.3.2 School-Based Model and Active Participation of Schools

It is important that the professional development workshops are arranged on a school basis rather than for the individual teacher account. Each school has the responsibility to create a friendly learning environment and develop the norm in which teachers can take it for granted to make changes and are willing to share their teaching and learning experiences with one another. Second, professional development should be organized in a long-term stance because teachers' professional growth takes time (Clarke and Hollingsworth 2002; Guskey 1986). Also of importance is the continuing focus on a specific theme for each school. Our experiences in implementing professional development for experienced mathematics teachers show that teachers' quantum-jump of professional growth has to take at least two or three years. Thus, the deficit model of one-shot workshops aiming at teacher mastery of prescribed skills and knowledge should be excluded.

Another key to LUSBP is that schools actively applied for the program, which reveals the intension of school teachers who expect to make changes to enhance their pedagogical power.

15.3.3 Cooperation of University Faculties in Taiwan

Another innovation of the program is to have about 90 % university faculties who have mathematics education expertise participate in LUSBP and lead the school-based professional development workshops.

15.4 Organization of LUSBP

LUSBP was initiated by the first author, the director of CCT. LUSBP started from 2012 and keeps running until now. Each semester, school teachers participating in LUSBP have to attend five or six sessions, each of which lasts for three hours. Table 15.1 is the summary of number of elementary and middle schools applied for LUSBP till the second semester of 2016 as well as the number of mathematics educators, most of which are MTE-Rs, involved in the project.

Those schools participated in LUSBP located in different municipalities and counties around Taiwan. The amount of schools and educators engaged in the program is an innovation in Taiwan mathematics education.

Another innovation for LUSBP is to include a variety of themes: conjecturing, modeling, reading comprehension, conceptual diagnostic teaching, inquiring, assessment for mathematics literacy, and teaching with DGS supports. As the first author has been the chair of Department of International Cooperation and Science

Table 15.1 Summary of schools participated in LUSBP

School year and semester	2013 Second semester	2013 First semester	2014 Second semester	2014 First semester	2015 Second semester	2015 First semester	2016 Second semester
Number of elementary schools	30	29	24	41	82	94	95
Number of middle schools	22	34	35	27	34	47	19
Sum	52	63	59	68	116	141	114
Number of educators	47	51	40	52	60	58	67

Education at Ministry of Science and Technology, he has well understanding of individual mathematics education faculty in terms of their research expertise around Taiwan. The understanding enables him to formulate the themes with the aim to well connect between research and practices. The underlying rationale for the themes is to create opportunities for Taiwanese students to learn mathematics in a more active way.

Before running LUSBP, we also invited experienced MTE-Rs to discuss the curriculum materials with correspondence to each theme and to structure the session agenda. We then open the program for schools and encourage them to select a theme which meets their school's need. Our experiences of implementing LUSBP for seven semesters also reveal that a high portion of schools who continued participating in LUSBP tended to change the learning themes from semester to semester. After schools determined their professional development themes, faculties who were mastered at or interested in the selected themes were assigned accordingly. Similar to our design-based workshops, LUSBP also asked teachers to initiate tasks, present their designs for obtaining feedback from educators and peer teachers, enact the designs with students, and revise the work accordingly. The whole process offered school teachers opportunities to detect and challenge their individual pedagogical problems.

Rather than evaluating LUSBP by quantitative report, each semester we hold a symposium and invite each school to share their learning and challenges.

15.5 The Learning of Students, Teachers, and Educators

The effectiveness and efficiency of professional development heavily relies on the quality of the interactions among tiers of participants including educators, teachers, and students. Of importance is that not only students but also teachers and educators can recognize their role as learners involved in the professional settings. Lesh and Kelly (2000), using mathematical modeling as an example, highlighted the importance of involving different tiers of participants who cooperate in an interactive nature to solve problems encountered in classrooms.

15.5.1 The Learning of Teachers and Students

The effectiveness and efficiency of individual school participating in LUSBP is carefully detailed in another chapter of this book, titled as "*School-based in-service mathematics teachers' professional development: The approach of diagnostic conjecturing activities designing*". Generally speaking, teachers' learning in LUSBP involves two aspects. One occurs in professional development workshops; and the other happens during the interactions with other school teachers and educators in symposiums. Teachers' change, especially the change in belief, usually involves a long-term journey. Although LUSBP schools are voluntary to apply the program, we still see a number of teachers have doubts of the values of the program at the beginning participation stage. For example, teachers may say that they do not know what they can learn from the program. Even though the teachers saw the differences between their teaching and others (e.g., student-centered teaching approach), they may still keep the doubts in mind. Those teachers usually are satisfied with their classroom teaching and think that they know their students well. One crucial turning point for the teachers to make change is to design and test learning activities (which contained tasks). Through designing and testing, teachers have opportunities to create or revise instructional activities that they have not implemented previously, thus creating opportunities for them to learn from their students. Another crucial turning point for the teachers can be the responses from their students, especially those they do not know previously.

There are two main aspects of teacher learning in this project. The first aspect involves teachers' learning from changing their classroom teaching for providing students more opportunities to think and construct mathematics actively. For example, teachers can try to change the teaching entry from posing a true propositions or concepts to false ones. Teachers can also change the teaching strategies that involve students engaging in enquiry-based learning activities. It is the importance of the change of teachers' roles from transmitter of knowledge to the facilitator of learning. The second aspect involves teachers' learning from enhancing the understanding of student learning. Teacher can learn from communication of students' specific performance. Teacher also can learn from reasoning students' learning difficulties and formulate the underlying reasons that constitute the difficulties. Also can be that teacher learn from connecting students' various responses.

The symposiums also offer teachers opportunities to learn from other school teachers and educators. Symposiums allow teachers to share the challenges encountered and discuss the possible strategies that can be used to overcome the challenges. Through participating in the LUSBP symposium, teachers can learn by observing how other schools implement the professional development workshops. They also can observe tasks designed by other teachers and then adopt and modify the tasks for their own use. Symposiums also allow teachers to discuss the effectiveness and efficiency of the ways of facilitating teacher learning so that the program can be improved accordingly.

15.5.2 The Learning of Educators

Educators' learning is another key of LUSBP. Educator's learning also plays crucial role of determining the effectiveness and efficiency of professional development programs. As Zaslavsky and Leikin (2004) stated that educators, similar to mathematics teachers, make professional growth through the practices involving research and implementation of professional development workshops with teachers.

More than that, we stated that it is crucial to create an environment where educators, including those university faculties, can learn from one another. To this end, we initiated a professional development forum for educators, most of which are MTE-Rs, after implementing LUSBP for a semester. This was due to the tensions encountered by educators when they facilitate teacher growth in professional development workshops. One tension has to do with the issue about utility of theory as teachers usually expressed that theory is not useful to their teaching practice. Another tension involves the identities of in-service teachers in relation to educators in professional development workshops. For example, some in-service teachers may attend workshops only for knowing if there's something new that they can directly use in their classroom. In this regard, they are passive receivers and usually refuse to involve in task design. The third is the tension related to teachers' vulnerability. Teachers may feel criticized and vulnerable when they receive suggestions and comments from peer teachers and the educators. In this situation, it is likely that teachers become too guarded to follow professional development agenda.

Those tensions described above became the discussion issues for the forum for the educators. The forum offered educators opportunities to express their perspectives related to the tensions and the ways that they dealt with them. Before educators discussed the issues in groups, some educators and experienced mathematics teachers were invited to share their experiences and perspectives. The sharing activities by the educators and experienced mathematics teachers helped other educators make sense of the tensions and provoke their memory of prior experiences related to the tensions. In addition to the discussions with the tensions, several educators, especially those experienced educators, were also invited to share how they implemented professional development with different themes initiated originally. The discussions among the educators and the presentation related to different themes were necessary to identify educators' initial, perhaps reflective, perspectives with respect to the tensions; which consequently affect their decisions and practices in facilitating teacher growth.

A total of forty-two educators, most of which are also faculties teaching at universities, participated in the forum. Most of the educators have Ph.D. degree in mathematics education. Some others major in mathematics and teach courses related to mathematics education at university level or have experiences with implementing professional development workshops for mathematics in-service teachers. Additionally, nine experienced mathematics teachers affiliated to CCT also participated in the forum. Those teachers allowed the educators to know more about the tensions from teachers' perspectives.

Taking an example involving the tension of elaborating theories and principles for participating teachers, we note that the tension is strongly related to teachers' perspectives on the utility of theory. We analyzed the data collected from the professional development workshop for educators and categorized educators' perspectives on utility of theory into three main types including concerns about meaning of theory; concerns about application of theory; and concerns about meta-coordination of theory. Details of teachers' perspectives on utility of theory can be seen in the forthcoming paper written by Lin et al. (In Submission).

15.6 Concluding Remark

Design is the core of solving educational problems; not only because it can create a diversity of curricular materials to enhance students' learning of mathematics but also a tool to facilitate professional growth of teachers and educators through the cycling process of creating tasks, enacting tasks with students, and then revising the tasks based on the reflection on student learning phenomenon. Another core for professional development is that educators, teachers, and students are all active learners; each of which takes his/her responsibility and work for his/her own learning. The ground-based and cooperation among three tiers of participants make it possible for the improvement of mathematics education in Taiwan and enable students with mathematics literacy and positive learning attitudes.

To this end, this chapter presents a nation-wide professional development program, namely LUSBP, in Taiwan that is established based on several innovations: (1) a relative high portion of schools and roughly 90 % of faculties who are mastered in mathematics education participate in the project; (2) school-based arrangement which bridges the gap between the research and practice as well as creates the environment where school teachers can learn from one another; (3) the implementation of teacher-as-designer approach; (4) enactment with a variety of themes under the core of students' active thinking; (5) the implementation of professional development workshop for educators to learn.

Our implementation of LUSBP for semesters shows that school-based professional development can facilitate teachers' growth through a design-based approach. Teachers' reflections highlight the value of the project as it can lighten up their minds and competences for better mathematics teaching and learning.

While recognizing the power of this project to teacher growth in profession, we also noticed the tensions faced by educators. As educators recognized the responsibilities to facilitate teachers' growth in profession, they encountered challenges and tensions due to teachers' different identities in the professional development program, the difficulty in elaborating theories and principles for teachers, and the vulnerable characteristics of teachers; which in turn may hinder the facilitation of teachers' professional growth. To solve those tensions encountered by educators, we held a professional development workshop where the participating educators can share their educative challenges and strategies with one another.

For teachers, they also face the challenges and tensions with respect to designing, teaching and students' learning. Designing involves teachers' confidence in creating the tasks, some teachers may think designing is the jobs of publishers but not theirs; thus refusing to participate in designing. The attitude can constrain those teachers' learning. Regarding the challenges related to teaching, it is about the struggle between content-oriented and student-centered. Taiwanese teachers usually concern how mathematical content can be successfully acquired by students instead of the ways that students participate in learning mathematics. As to students' learning, teaching for different levels/populations of students to promote their active thinking is the key for consideration. Another tension is about how to structure the tasks and manage the openness of the problems so that students know how to do it in an active way. Those challenges and tensions require long-term cooperation between educators and teachers in a reciprocal relationship for self-understanding and re-conceptualization of mathematics teaching and professional development (Jaworski 2001).

References

Bishop, A. J. (1991). *Mathematical enculturation: A cultural perspective on mathematics education*. Dordrecht: Kluwer Academic Publishers.

Clarke, D., & Hollingsworth, H. (2002). Elaborating a model of teacher professional growth. *Teaching and Teacher Education, 18*, 947–967.

Claxton, G. (2002). Learning to learn: A key goal in a 21st century curriculum. *Futures meeting the challenges. Qualifications and Curriculum Authority paper.*

Cobb, P., Confrey, J., diSessa, A., Lehrer, R., & Schauble, L. (2003). Design experiments in educational research. *Educational Researcher, 32*(1), 9–13.

Cooney, T. J. (1994). On the application of science to teaching and teacher education. In R. Biehler, R. W. Scholz, R. Straber, & B. Winkelmann (Eds.), *Didactics of mathematics as a scientific discipline*. New York: Kluwer Academic Publishers.

Freudenthal, H. (1983). *Didactical phenomenology of mathematical structures*. Dordrecht: D. Reidel.

Gravemeijer, K. (1994). Educational development and developmental research in mathematics education. *Journal for Research in Mathematics Education, 25*(5), 443–471.

Guskey, T. R. (1986). Staff development and the process of teacher change. *Educational Researcher, 15*(5), 5–12.

Jaworski, B. (1993). The professional development of teachers—The potential of critical reflection. *British Journal of In-Service Education, 19*(3), 37–42. doi:10.1080/0305763930190307

Jaworski, B. (2001). Developing mathematics teaching: Teachers, teacher educators, and researchers as co-learners. In F.-L. Lin & T. J. Cooney (Eds.), *Making sense of mathematics teacher education* (pp. 295–320). Dordrecht: Kluwer Academic Publishers.

Lakatos, I. (1976). *Proof and refutations: The logic of mathematical discovery*. Cambridge: The Cambridge University Press.

Lesh, R. (2003). Research design in mathematics education: Focusing on design experiments. In L. D. English (Ed.), *Handbook of international research in mathematics education* (pp. 27–49). Mahwah, New Jersey: Lawrence Erlbaum Associates.

Lesh, R., Hamilton, E., & Kaput, J. (2007). Directions for future research. In R. Lesh, E. Hamilton, & J. Kaput (Eds.), *Foundations for the future in mathematics education* (pp. 449–454). Mahwah: Lawrence Erlbaum Associates.

Lesh, R., & Kelly, A. (2000). Multitiered teaching experiments. In A. E. Kelly & R. Lesh (Eds.), *Handbook of research design in mathematics and science education* (pp. 197–230). Mahwah, New Jersey: Lawrence Erlbaum Associates.

Lin, F.-L. (2009). "Harmony, east attainment and thoughtfulness"—Teaching and learning in mathematics. In *Paper Presented at the Seminar on Mathematics Education-Theory and Perspective of Mathematics Learning and Theory from the East*, Hong Kong.

Lin, F.-L., & Tsao, L.-C. (1999). Exam math re-examined. In C. Hoyles, C. Morgan, & G. Woodhouse (Eds.), *Rethinking mathematics curriculum* (pp. 228–239). London: Falmer Press.

Lin, F.-L., Yang, K.-L., Hsu, H.-Y., & Chen, J.-C. (In Submission). Utility of theory in facilitating teacher growth: Mathematics teacher educator-researcher perspectives. *Educational Studies in Mathematics*.

Michener, E. R. (1978). Understanding mathematics. *Cognitive Science, 2*, 361–383.

Ministry of Education. (2008). *Establishment and operation of central curriculum and instruction counseling team*. Taipei, Taiwan: Ministry of Education.

Mullis, I. V. S., Martin, M. O., & Foy, P. (2008). *TIMSS 2007 international mathematics report*. Chestnut Hill, MA: TIMSS & PIRLS International Study Center, Boston College.

Mullis, I. V. S., Martin, M. O., Foy, P., & Arora, A. (2012). *TIMSS 2011 international results in mathematics*. Chestnut Hill, MA: TIMSS & PIRLS International Study Center.

OECD. (2005). *PISA 2003 technical report*. Paris: Authors.

Reid, D., & Zack, V. (2010). Observing the process of mathematics teacher change-part 1. *Journal of Mathematics Teacher Education, 13*(5), 371–374.

Ruthven, K., Laborde, C., Leach, J., & Tiberghien, A. (2009). Design tools in didactical research: Instrumenting the epistemological and cognitive aspects of the design of teaching sequences. *Educational Researcher, 38*(5), 329–342.

Smyth, J. (1989). Developing and sustaining critical reflection in teacher education. *Journal of teacher education, 40*(2), 2.

Yang, K.-L., Hsu, H.-Y., Lin, F.-L., Chen, J.-C., & Cheng, Y.-H. (2015). Exploring the educative power of an experienced mathematics teacher educator-researcher. *Educational Studies in Mathematics, 89*(1), 19–39. doi:10.1007/s10649-014-9589-4

Zaslavsky, O., & Leikin, R. (2004). Professional development of mathematics teacher educators: Growth through practice. *Journal of Mathematics Teacher Education, 7*, 5–32.

Chapter 16
School-Based In-service Mathematics Teachers' Professional Development: Designing Diagnostic Conjecturing Activities

Jian-Cheng Chen and Fou-Lai Lin

Abstract This chapter reports the professional learning of 21 in-service mathematics teachers (MTs), from two junior high schools participating, in the Lighten-Up School-Based Program (LUSBP). The MTs in both the schools chose mathematical conjecturing as the theme of professional development (PD); they also participated in the PD workshops on approaches for designing diagnostic conjecturing activities planned and implemented by the same mathematics teacher educator–researcher (MTE-R). The data were collected from the dialogs and textual materials of the participating MTs, MTE-R, and students. We analyzed these data from three dimensions: the teaching designs, teaching experimentations, and environments of professional development. The results indicated the MTs' changes included: be oriented to student-centered teaching approach, be more sensitive to students' learning, and established the school-based professional learning communities. According to the above results, we proposed suggestions to the plan and implementation of future professional development courses for in-service mathematics teachers.

Keywords Diagnostic conjecturing · Tasks designing · School-based · Mathematics teacher · Teacher professional development

16.1 Introduction

In Taiwanese examination-driven culture, the junior high school mathematics teachers were mostly oriented to teacher-centered instruction (talk and chalk), they tried hard to present as many examples, including the examination questions,

J.-C. Chen (✉)
Ming Chi University of Technology, New Taipei City, Taiwan
e-mail: janchen@mail.mcut.edu.tw

F.-L. Lin
National Taiwan Normal University, Taipei City, Taiwan
e-mail: linfl@math.ntnu.edu.tw

B. Kaur et al. (eds.), *Professional Development of Mathematics Teachers*, Mathematics Education – An Asian Perspective,
DOI 10.1007/978-981-10-2598-3_16

elegant solutions, and memorable strategies, as possible in limited time and expected students to obtain higher grades in the examinations (Lin and Tsao 1999). For long, such a teacher-centered instruction was difficult to be changed. To this end, the Ministry of Education's (MoE) National Team for Educating In-service Teachers has promoted the Lighten-Up School-Based Program, aiming to establish the school-based professional learning communities for teachers as well as plan the professional development courses on approaches for designing teaching and learning activities (Lin 2012). The rationale of this program demanded that the participating MTE-Rs, MTs, and students were active learners. The program planned seven innovative teaching themes (e.g., mathematical conjecturing, diagnostic teaching, and reading comprehension, etc.) for the MTs of participating schools to choose, it also matched appropriate MTE-Rs to visit their corresponding schools so as to collaborate with in-service MTs in PD workshops. Through a series of innovative teaching designs and experimentations, the participating MTs gradually learned and were inclined to learner-centered teaching approach. This chapter presented a case of teachers' professional development on approaches for designing diagnostic conjecturing activities.

16.2 Theoretical Background

In order to investigate MTs' learning processes and results in the design-based PD workshops, the perspective of a reflective practitioner was adopted (Zaslavsky and Leikin 2004), which meant the learning of MTs referred to their reflection upon their own designing and experimental practices. This chapter emphasized that MTs could develop certain strategies to facilitate students' learning performances as well as further enhance the effects of these strategies and their own professional learning through the processes of designing, experimenting, and reviewing. Furthermore, we particularly highlighted the designing of innovative teaching activities which the participating MTs would like to learn: the diagnostic conjecturing activities, integrating diagnostic teaching with conjecturing approach.

16.2.1 MTs' Design-Based Professional Development

Clarke and Hollingsworth (2002) proposed The Interconnected Model of Teacher Professional Growth made up of four domains and two mediating mechanisms (e.g., reflection and enactment), which was used to describe and identify teachers' change sequences and growth networks as well as emphasize on the importance of the support from a school context. The LUSBP in Taiwan adopted their approach to facilitate MTs' PD but stressed on: (1) the school-based learning communities: MTs in each school served as the main body of PD who actively chose their required

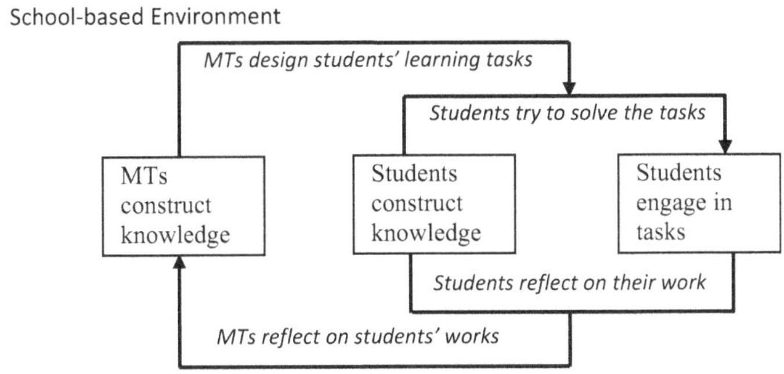

Fig. 16.1 MTs' design-based PD in school-based workshops

teaching approach, with the appropriate MTE-Rs matched by the program to visit these MTs in their schools; (2) the approach of design-based PD: MTs were the designers of teaching activities, the instructors of teaching experimentations, and the reflective practitioners of experimental results. Meanwhile, these MTs continuously improved the effects of their designing; (3) the perspective of participants as learners: in the program, the participating MTE-Rs, MTs, and students were all learners, who facilitated their mutual learning through three levels of interactions. To show these features, we adapted Zaslavsky's Model of Facilitator-Learner (2008) for the MTs' professional development framework as shown in Fig. 16.1.

Within this framework, we could divide teachers' PD into two dimensions: results and processes. As for the results of PD, the MTs facilitated their own construction of professional knowledge through reflective practice (Schön 1983). Ideally, the MTs' constructed knowledge should at least contain these domains: mathematics, mathematics learning, and mathematics teaching, etc. (Jaworski 2001; Shulman 1986). Moreover, MTs would have sufficient opportunities for the construction of shared knowledge, including professional knowledge and scientific knowledge (Jaworski 2008). With regard to the processes of PD, the processes of MTs' designs and experimentations of innovative teaching could achieve a twofold aim: on the one hand, the MTs' designs of teaching activities and their teaching experimentations based on the designed activities could facilitate and improve students' learning performances; on the other hand, the MTs' processes of examining and revising teaching activities, and their continuous reflection and enactment also facilitated their own professional growth (Zaslavsky 2008). If MTs could integrate professional development with action research (Kemmis and McTaggart 1988) to continuously improve their designs and experimentations of teaching activities; moreover, with the intervention of research provided more opportunities for critical reflection, which should help improve the quality of knowledge construction (Jaworski 2006). Due to the word limitation, this chapter only focused on the results rather than the dynamic processes of PD in school-based workshops.

16.2.2 Design of Diagnostic Conjecturing Activities

There were many literatures concerned about mathematical conjecturing, which included: (1) the meanings and functions of mathematical conjecturing, e.g., Mason et al. (1982) applied various mathematical cases to elaborate the meanings and functions of conjecturing and highlighted conjecturing was the backbone of problem-solving; (2) the cognitive types of mathematical conjecturing, e.g., Cañadas et al. (2007) elaborated the cognitive types and processes of five conjecturing, including induction from a finite number of discrete cases and from dynamic cases, analogy, abduction as well as perceptually based conjecturing; (3) the designing strategies of mathematical conjecturing tasks, e.g., Lin (2006) proposed three entries of conjecturing activities designing and their corresponding strategies, including false proposition as entry with the proceduralized refutation model (Lin et al. 2005), true proposition as entry with the strategy of analogies or what if not and others, and students' conjecture as entry with the strategy of defining or constructing premise/conclusion; (4) the designing principles for mathematical conjecturing tasks, e.g., Lin et al. (2012) proposed four designing principles for conjecturing tasks, which included providing an opportunity to engage in observation, to engage in construction, to transform prior knowledge, and for reflection. The learning opportunities described by these four principles would facilitate making conjectures, which could be further applied in the evaluation of conjecturing tasks designed by the MTs (Lin et al. 2014).

Refuting was one important activity of mathematical conjecturing and argumentation, mathematicians often applied it to understand and examine propositions (Lakatos 1976). Lin (2005) argued that refuting could facilitate students' critical thinking. In particular, false proposition as entry of learning activities could motivate students to conduct refuting: upon facing false propositions, students would generate supportive and counterexamples based on the given propositions, observe the common properties among the supportive examples, make conjectures based on the observing properties, or convince oneself to the correctness of conjectures, etc. On these grounds, Lin et al. (2005) developed the proceduralized refutation model (PRM), referring to the processes of mathematicians' refuting and transforming them into the teaching activity sequences. With the focus of MTs as guides, MTs' timely intervention would engage students' participation as well as enhance their competencies in refuting and conjecturing. Our experiences showed that Taiwanese in-service MTs seemed to be more sensitive to students' conceptual understanding than the approach of conjecturing and argumentation. To design an attractive entry for MTs' PD program, Chen et al. (2014) developed the theoretical perspective of the integration of conceptual diagnostic teaching and conjecturing activities on the basis of the PRM; we also proposed the framework for designing diagnostic conjecturing activities as shown in the following table. The framework could be used to generate diagnostic conjecturing activities; moreover, the results of empirical studies showed these kinds of teaching activities functioned as the facilitation of conceptual understanding and making conjectures.

16.3 Research Methods

The research methods section was divided into two subsections: research contexts and data analysis. The first subsection explained the subjects, the fields of professional development, and the research methods; the second was about data collection and data analysis.

16.3.1 Research Contexts

The research subjects were 8 and 13 MTs with 15 years of average teaching experience in two junior high schools, respectively. These two schools applied for the LUSBP actively and both chose diagnostic conjecturing activities as the approach of their professional development; the first author of this chapter was matched by the LUSBP to serve as the MTE-R who provided mentoring services there. The MTE-R planned the PD program for six times of visiting to each school, each time lasting for 3 hours as a course over a semester. Responding to MTs' multiple roles, the main activities included the following: (1) entry activity: MTs were the teaching practice experts who commented on the examples of diagnostic conjecturing activities (which could also be called generic examples for activities designing, referring to Appendix 1) to draw out MTs' professional viewpoints and promote the communication of various practical and theoretical viewpoints, especially the MTE-R's intervention of using the PRM as a basis to develop the theoretical perspective of the integration of diagnostic teaching and conjecturing activities (Chen et al. 2014); (2) designing and experimental activities: MTs were the designers of conjecturing activities who designed conjecturing activities using theoretical perspectives as the bases and referring to the examples of diagnostic conjecturing activities and then conducted revisions and adjustments on the bases of thinking experimentations; meanwhile, MTs were the instructors of teaching experimentations who applied their designed activities in classrooms and conducted revisions and adjustments on the bases of their teaching experimentations; (3) reviewing activity: MTs were the reflective practitioners who could carry out critical reflection and adjustments during and after the processes of PD workshops, particularly focusing on the argumentations of the relation between the results of their teaching experimentations and the intentions of their teaching designs; the relation of these three activities was shown in Fig. 16.2.

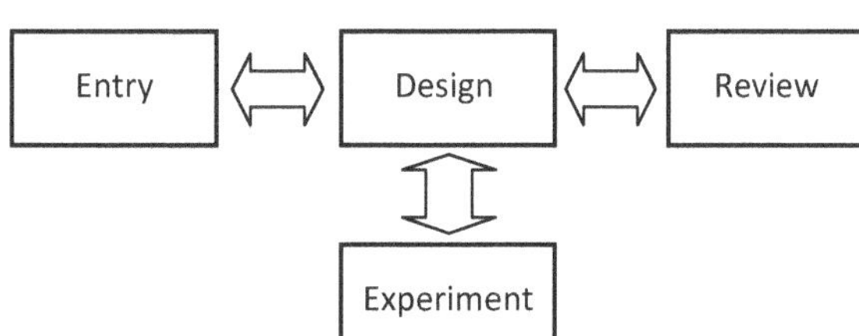

Fig. 16.2 MTs' PD activities on approaches for designing diagnostic conjecturing activities

16.3.2 Data Processing

To understand MTs' performances in design-based workshops, the data were collected from: video- and audio-recordings from each of the workshops; textual materials including MTs' individual comments on the conjecturing products, the designed products of conjecturing activities, students' performances in teaching experimentations, and MTs' reflective reports; the videos and PPTs of school representatives shared in the midterm outcome reports; MTE-R's presentation materials during the workshops, observational notes, and reflective reports. The various data sources were used to enrich and correct MTs' performances.

In order to analyze MTs' professional learning and changes on approaches for designing diagnostic conjecturing activities, the data analysis section divided the theoretical background into two levels. First, we identified three dimensions of teachers' changes in accord with the framework for MTs' PD in Fig. 16.1: (1) designing dimension: examining MTs' enactment and reflection of designing teaching activities which included conceptualization, teaching designs, and thinking experimentations; (2) experimental dimension: examining MTs' enactment and reflection of conducting teaching activities which included teaching experimentations, analysis of students' performances, as well as assessment and revision; (3) environmental dimension: examining MTs' professional interactions and reflection in school-based learning environments which included MTs' dialogs with other MTs, MTE-R, or school principal, etc. Second, according to the framework for designing diagnostic conjecturing activities in Table 16.1, we extracted some important elements from the teaching activities such as entry and approach. The perspectives to examine MTs' entry and approach of designing teaching activities included knowledge, beliefs, and attitudes, etc. These data were used to show MTs' professional development in design-based workshops and generalize the trends of MTs' changes.

16.4 MTs' Professional Learning

According to the results of data analysis with regard to the teaching designs, teaching experimentations, and environments of professional development, we presented MTs' changes in three dimensions: teaching orientation, sensitivity to students' learning, and establishment of school-based professional learning communities, respectively.

16.4.1 MTs' Changes of Teaching Orientation

Through designing and experimenting diagnostic conjecturing activities, MTs were aware of misconceptions as teaching resources, experienced teaching with students'

Table 16.1 Framework for designing diagnostic conjecturing activities

	Features
Learning theme	Teaching material content
Learning aim	To facilitate conceptual understanding and conjecturing thinking
Conjecturing entry	Students' possible misconceptions (false propositions)
Cognitive type	Conjecturing by induction from cases
Learning strategy	Proceduralized refutation model
Activity sequence	1. Students observe a mathematics situation that can elicit their misconceptions 2. Students generate both supportive and counterexamples based on the observation of the mathematics situation 3. Students identify common properties embedded in the supportive examples 4. Students revise the mathematics situation and provide correct statements 5. Students justify the revised mathematics situation
Expectation	Anticipation of students' performance in teaching experimentation

inquiry-based learning approach, and recognized teachers as facilitators; all these changes seemed to be oriented to learner-centered teaching approach (Weimer 2013).

(1) Be aware of misconceptions as teaching resources: MTs were aware that students' common misconceptions were important teaching resources which could also be taken as entries of conjecturing activities and the core of diagnostic teaching approach, with the aim to avoid students creating these misconceptions before learning rather than traditionally to correct students' formed misconceptions after learning. At the beginning, most MTs believed false proposition (or misconception) should not be taken as the entry of teaching new concepts. For instance: teacher TY01 believed MTs should provide true proposition or example as entry to avoid students' confusion in conceptual learning; he said, "*If I were the teacher, I would provide obtuse-angled isosceles triangle, right-angled isosceles triangle, and acute-angled isosceles triangle…, let the students discover the key factor is about being 'isosceles' while ruling out other factors, without too many variables, or the focus would be blurred.*" After participating in the workshops and by the designs and experimentations of diagnostic conjecturing activities, most MTs were aware that misconception as entry would brought learning opportunities for students. For instance: teacher TY01 indicated, "*this allowed me to know that in designing a learning activity, we can first take misconception as entry, and then let students generate examples, observe properties, make conjectures, and create proofs.*" Moreover, most MTs could design teaching activities on

the bases of students' misconceptions, which was selected by them, e.g., according to the past experiences in teaching multiplication formula, teacher TX01 took the students' common misconception $(a+b)^2 = a^2 + b^2$ as entry to design his teaching activity, as shown in Appendix 2.

(2) Experience teaching with students' inquiry-based learning approach: MTs' teaching approach changed from teachers' instruction and notification to students' exploration and discovery. At the beginning, the MTs set up their teaching goal for students to obtain knowledge, they believed the most effective teaching approach was to instruct knowledge rapidly and clearly. After participating in the workshops and by the designs and experimentations of diagnostic conjecturing activities, they experienced the teaching with students' inquiry-based learning approach and the impact brought by this approach. For instance: teacher TY01 stated in his reflective report, "*by this kind of activity, students could explore, discover, and generalize mathematical conclusions by themselves; meanwhile, they could correct certain past misconceptions… even low achievers could approximately exactly complete this learning activity, which achieved the learning effect and also became the largest harvest for my participation in this workshop!*" For most MTs, taking students' inquiry-based learning as their teaching approach was still unfamiliar for them. Nevertheless, most participating MTs were willing to undergo these processes, e.g., teacher TX01 conducted three times of teaching experimentations during the workshops to continually revise his way of intervention and grasp his perspective of implementing this teaching approach, he said, "*the learning activities should be arranged and presented in sections with respect to five parts of designing conjecturing activities (leaving time for students to discuss and share) and the inquiries should be in sequences, guiding students to think actively seems to be the more ideal teaching approach.*"

(3) Recognize teachers as facilitators: MTs' role changed from the transmitter of mathematical knowledge to the facilitator of students' inquiry-based learning; meanwhile, students' role also changed from the receiver of knowledge to the constructor of learning. After participating in the workshops and by the designs and experimentations of diagnostic conjecturing activities, they recognized the changes of teachers' and students' roles brought differences in teaching and learning. For instance: teacher TY06 stated in his reflective report, "*what students learned was to actively generate more examples to challenge the rules and properties, to make reasonable conjectures (even if these conjectures are false, the formation and revision of conjectures still provided sufficient opportunities for students to think), to experience the progressive processes from incomplete knowledge to complete knowledge. This kind of mathematics class is a whole new world for me.*" For many MTs, though the teaching designs and experimentations caused the changes of teachers' and students' roles, the teacher–student interaction still required to be further established. For instance: the first time when teacher TX01 let the students learn in collaborative group work, he was aware that the atmosphere of students' collaborative

learning was not formed; furthermore, he also lacked of appropriate way of intervention which made this learning activity lose focus. Thus, he declared the transformation of different roles would take a long time, he said, "*running a course in collaborative group work needs to be conducted chronically so that it would become well-regulated.*"

16.4.2 MTs' Sensitivity to Students' Learning

Through designing and experimenting diagnostic conjecturing activities, MTs were able to be aware of students' remarkable learning performances, analyze students' learning difficulties and possible causes as well as draw up corresponding solutions, and recognize unimaginable students' zone of proximal development (ZPD) (Vygotsky 1978); the above results indicated MTs' sensitivity to students' learning.

(1) Be aware of students' remarkable learning performances: when MTs reported students' remarkable performances in the experimentations, especially the interactions between MTs and students; it showed MTs' sensitivity and understanding to students' mathematical power. For instance, teacher TX03 proposed one of students' common misconceptions in one workshop, the converse of "*proportional segments theorem: If three lines cut two transversals into proportional segments, then these three lines are parallel;*" although the MT could give examples to illustrate the proposition was false, he had not thought of and could not revise the premise of this proposition to make a true proposition. In the next workshop, he reported his interactions with a certain student after posing this question; their interactions were excerpted as follows:

T *If three lines cut two transversals into proportional segments, are these three lines parallel?*

S (*Using a ruler to draw many figural examples as shown in* Fig. 16.3a)..., *the statement is false, linking proportional segments wouldn't become parallel lines...*

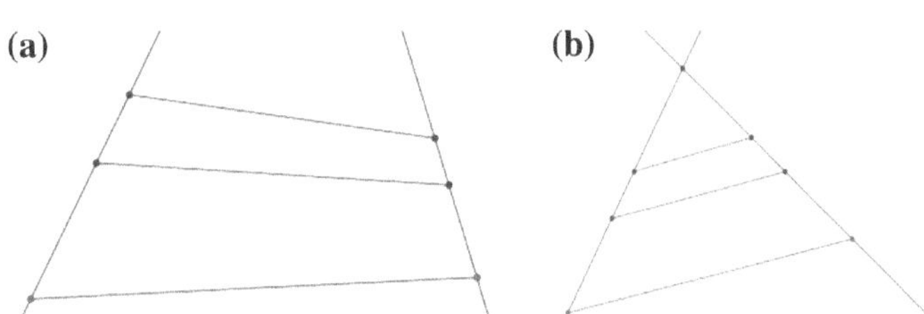

Fig. 16.3 **a** The supportive example generated by student. **b** The counterexample generated by student

T *There might be some occasions that would make parallel lines; you need to add certain conditions...*

S *...(After one class, the student actively reported his finding to the MT), if extending two transversals to be intersected at one point (to form a triangle as shown in* Fig. 16.3b) *and the proportional segments are on the basis of the intersection point, then these three lines would become mutually parallel...*

T *(Be surprised at the student's performance), applying geometric properties to prove the proposition revised by the student is true.*

This student's performance impacted on teacher TX03; at that time, he were aware that students' mathematical power could even surpass the teachers' and teachers could learn from students, this also facilitated MTs' deeper sensitivity and understanding to the specific false proposition (or misconception) regardless of mathematics and students' learning. In mathematics, MTs realized the revision of the premise or conclusion of a false proposition could make a true proposition; in students' learning, MTs understood misconception as entry could provide opportunities for students to learn actively and stimulate their learning power.

(2) Analyze students' difficulties and causes: MTs analyzed students' learning difficulties, reasoned, and conjectured possible causes of these difficulties, and further drew up corresponding solutions; this also showed their sensitivity and understanding to students' mathematical power. For example: teacher TX01 reported two classes of seventh graders' performances in generating examples, there were 12/17 groups provided examples with negative numbers and 10/12 among them made the same errors. Some students took $a = -1$ and $b = -4$ as an example to calculate the sums of $(a + b)^2$ and $a^2 + b^2$; they obtained the sums as 25 and −17, respectively; the previous one was positive while the latter was negative. Teacher TX01 then conjectured this was because the students were unfamiliar with the symbolic computation and suggested postponing the teaching experimentations. He said, "*because the students have yet learned to substituting symbols with numbers which causes a series of calculating errors..., it is suggested arranging the teaching experimentations in the second semester of 7th grade.*" Students' common errors attracted participating MTs' attention and heated discussions, teacher TX08 conjectured that one possible cause of students' errors was due to their confusion about the order of operations of various symbols and thus suggested using brackets to help students correctly generate the examples, he said, "*7th graders just start to learn negative numbers, they are easily to have the confusion about the order of operations..., we stressed to deal the formulas with brackets first in class, and students' performances in those with brackets are correct..., I suggest adding brackets to* $a^2 + b^2$ *as* $(a)^2 + (b)^2$." In sum, the impact which MTs encountered in students' learning difficulties during their teaching experimentations enhanced MTs' deeper sensitivity and understanding to students' learning problems, possible causes, and corresponding solutions.

(3) Recognize unimaginable students' ZPD: from students' performances in the experimentations, MTs recognized students' ZPD might beyond their imagination. For instance, teacher TX05 took students' misconception "*in any triangle, the median on the base is the altitude of the triangle*" as entry to guide eighth graders conduct cognitive activities such as generating examples, observing common properties, revising propositions, reasoning and argumentation, etc. From students' performances in the teaching experimentations, he recognized unimaginable students' mathematical power and attitudes. He addressed, "*although students are unable to conduct geometrical argumentation, they are able to use the terms or concepts such as congruence, symmetry, or perpendicular bisector which will be taught until 9th grade…; the average degree students feel more confident, they feel the new unit (cross-product method) is much easier…; most students think this kind of teaching activities is good, but it takes too much time and energy and should not be conducted too often.*" In addition, teacher TX05 found unimaginable students' performances would stress their ZPD, he said, "*one student could argue the correctness of proposition by Pythagorean Theorem; one student found the method to make regular triangles in graph paper….*" He even sighed that the examination system restrained students' mathematical power, he said, "*there's one student who only got 3-score in the term examination, but he is the first one who could draw the example of regular triangle and induced the conclusion exactly.*" The MT was impressed by the mathematical power students showed, but he also pointed out the limits and difficulties in keep on promoting this kind of learning activities; he addressed, "*conjecturing activities provide various learning approaches for students and their responses are positive and beneficial, but this kind of activities couldn't last long when facing entrance examinations. It is good to have the activity once in a while, but the most difficult part is to design the learning activities, particularly those matching to the contents of textbooks.*" The impact caused from unimaginable students' ZPD would enhance MTs' deeper sensitivity and understanding to students' learning opportunities and mathematical power with respect to the diagnostic conjecturing activities.

16.4.3 MTs' Establishment of School-Based Professional Learning Communities

For the MTs in these two schools, their past meetings used to discuss the administrative affairs in schools; however, after participating in the school-based PD workshops, they just had the opportunities to share and discuss their mutual teaching perspectives and students' in-class performances so as to facilitate the professional development of themselves and their learning communities, including (1) Share professional knowledge together: through the workshops, the professional knowledge that MTs as teaching practice experts could be shared mutually and further facilitate their professional learning and development, e.g., teacher TY09 indicated the interactions between colleagues served as one source for learning, he indicated, "*...by the design of learning activities as well as the discussions and interchanges with the MTE-R and colleagues, which enhance my professional development and teaching experiences invisibly....*" In addition, teacher TY12 emphasized colleagues' mutual learning also included attitudes, he said, "*during the PD workshops, I noticed the efforts of my colleagues; I really appreciated the support, comments, and suggestions offered by them and the MTE-R; this is the so-called teaching and learning promote and enhance each other.*" (2) Develop the professional learning communities: besides sharing professional knowledge mutually in the workshops, the MTs supported and encouraged one another in schools, this also stimulated the learning atmosphere of their communities which tended to be active participation, just as what teacher TY10 stressed in his outcome report, "*when we first attended the lessons, we treated everything with great concern, quietly immersed ourselves in hard work, very attentive, and sure to complete the tasks; now when we attend the lessons, we are engaged in heated discussions, and even after a lesson, we still won't let the teacher (MTE-R) go.*"

After participating in the PD workshops on approaches for designing diagnostic conjecturing activities, the MTs in these two schools still actively applied for the LUSBP, which indicated the school-based learning communities were successfully established and also showed quite different professional learning. For example, one successive MTE-R found the learning attitudes of the MTs in these schools (communities) were quite aggressive, she addressed, "*due to certain reason, one MT couldn't attend a certain workshop, he asked his colleagues to videotape the process; he then watched the video after work, completed the assignment at that time, and shared his outcome in the next workshop.*" Meanwhile, it also indicated that the MTs in these schools (communities) were aware that innovative teaching designs tended to be student-centered, she stated, "*once I offered my suggestions to a certain MT, he immediately replied he understand what I meant was expecting them to design student-centered teaching activities.*"

16.5 Reflection

This chapter proves that MTs' participation in professional development workshops on approaches for designing diagnostic conjecturing activities could facilitate their PD, provide opportunities for MTs to establish school-based learning communities, enhance MTs' sensitivity and understanding to students' learning, and change MTs' teaching orientation; no matter their teaching entry, approach, and teachers' role were gradually changed to student-centered teaching. Nevertheless, some empirical studies also proved these changes were long term and slow (Weimer 2013). Whether MTs would continuously show these changes in their teaching practices and completely change to student-centered teaching approach after the workshops would not be inferred and discussed in this chapter. Within the school-based PD workshops, the power which stimulated MTs' professional development came from three levels: (1) MTE-Rs: especially the MTE-R's intervention of framework and examples for designing teaching activities based on the integration of theories and practices; (2) students: especially their performances stimulated by the participation of innovative teaching experimentations; (3) colleagues: especially the establishment of school-based professional learning communities, where they could support and challenge one another. The above would be considered as important references to the plan and implementation of future professional development courses for in-service teachers.

Appendix 1: Diagnostic Conjecturing Activity Case

Class ______ No.: _______Name:________

Xiao-Ming has drawn a triangle and found the midpoint of the triangle's longest side for drawing a circle, which just connects the three vertices of the triangle; he then says: "in a triangle, we can use the midpoint of its longest side for drawing a circle that just connects all its vertices."

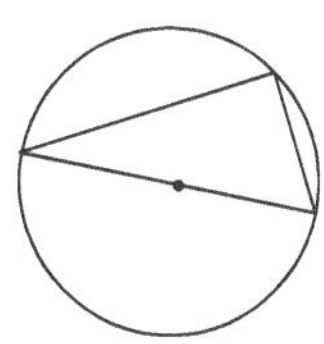

Is Xian-Ming correct in saying this? Let us validate this.

1. Please use the computer drawing program to construct various triangles and find their longest sides, and then use the midpoints for drawing circles and record the length of their sides and sizes of their angles in the following table:

Number	Length			Size			Name of triangle
	Side 1	Side 2	Side 3	Angle 1	Angle 2	Angle 3	

2. According to your records above, **which are the triangles that have the midpoints of their longest side which can be used for drawing a circle to just connect their three vertices**?
   - Please write out their numbers.
   - Say what common properties of these triangles have.
3. Please correct what Xiao-Ming has found:
   In ____________ triangle, we can use the midpoint of the longest side as the center for constructing a circle that just connects the three vertices.
4. If someone believes that the statement in 3. is only correct for the circle you have constructed but there is no guarantee that this is also correct for any other circles not yet constructed, can you give a reason that this statement must be correct? How do you say it?

Appendix 2: Diagnostic Conjecturing Activity Case

Learning sheet for mathematical conjecturing activity
Topic: Multiplication Formula Activity Duration: Class:

Members of group__: (1) No.:__ Name:____; (2) No.:__ Name:
(3) No.:__ Name:____; (4) No.: __Name:

[Question]
Xiao-Ming told Xiao-Hua: "Is 'the square of the sum of two numbers' equal to 'the sum of the squares of these two numbers'?"
Xiao-Hua replied: "It should be!"
Xiao-Juan Said: "I don't think so!"
Which statement is correct? Let's keep reading....

(1) Compare the number values: Please fill any numbers in a and b, and then complete the following table.
[Note]: a, b need to combine positive number, negative number, and zero; do not choose large numbers to avoid calculating error.

a	b	$a+b$	$(a+b)^2$	a^2	b^2	a^2+b^2	$(a+b)^2-(a^2+b^2)$

(2) According to the records above, what do you think of the relative magnitude of $(a+b)^2$ and (a^2+b^2) …

(3) According to the value of $(a+b)^2-(a^2+b^2)$, what do you think of its relation with a and b…

References

Cañadas, M., Deulofeu, J., Figueiras, L., Reid, D., & Yevdokimov, O. (2007). The conjecturing process: Perspectives in theory and implications in practice. *Journal of Teaching and Learning, 5*(1), 55–72.

Chen, J.-C., Lin, F.-L., Hsu, H.-Y., & Cheng, Y.-H. (2014). *Integration of conjecturing and diagnostic teaching: Using proceduralized refutation model as intermediate framework.* Paper presented at the the Joint Meeting of PME 38 and PME-NA 36, Vancouver, Canada: PME.

Clarke, D., & Hollingsworth, H. (2002). Elaborating a model of teacher professional growth. *Teaching and Teacher Education, 18*, 947–967.

Jaworski, B. (2001). Developing mathematics teaching: Teachers, teacher educators, and researchers as co-learners. In F.-L. Lin & T. J. Cooney (Eds.), *Making sense of mathematics teacher education* (pp. 295–320). Dordrecht: Kluwer Academic Publishers.

Jaworski, B. (2006). Theory and practice in mathematics teaching development: Critical inquiry as a mode of learning in teaching. *Journal of Mathematics Teacher Education, 9*(2), 187–211. doi:10.1007/s10857-005-1223-z

Jaworski, B. (2008). Mathematics teacher educator learning and development. In B. Jaworski & T. Wood (Eds.), *The international handbook of mathematics teacher education Volume 4: The mathematics teacher educator as a developing professional* (pp. 335–361). Rotterdam: SensePublishers.

Kemmis, S., & McTaggart, R. (1988). *The action research planner*: Geelong, Victoria: Deakin University Press.

Lakatos, I. (1976). *Proof and refutations: The logic of mathematical discovery.* Cambridge: The Cambridge University Press.

Lin, F.-L. (2005). *Modeling students' learning on mathematical proof and refutation.* Paper presented at the Proceedings of the 29th Conference of the International Group for the Psychology of Mathematics Education, Melbourne.

Lin, F.-L. (2006, January). *Designing mathematics conjecturing activities to foster thinking and constructing actively.* Paper presented at the Keynote address in the APEC-TSUKUBA International Conference, Japan.

Lin, F.-L. (2012). The criteria for the implementation of 101 academic year Lighten-Up School-Based Program. Taipei, Taiwan. Retrieved June 1, 2013, from http://tame.tw/highlight/forum.php?mod=viewthread&tid=3&extra=page%3D1

Lin, F.-L., & Tsao, L.-C. (1999). Exam math re-examined. In C. Hoyles, C. Morgan, & G. Woodhouse (Eds.), *Rethinking mathematics curriculum* (pp. 228–239). London: Falmer Press.

Lin, F.-L., & Wu Yu, J.-Y. (2005, July). *False proposition as a means for making conjectures in mathematics classrooms.* Paper presented at the Invited speech in Asian Mathematical Conference, Singapore.

Lin, F.-L., Yang, K.-L., Hsu, H.-Y., & Chen, J.-C. (2014). *Using theoretical principles in the evaluation of teachers' task design for conjecturing.* Paper presented at the Joint Meeting of PME 38 and PME-NA 36, Vancouver, Canada: PME.

Lin, F.-L., Yang, K.-L., Lee, K.-H., Tabach, M., & Stylianides, G. (2012). Task designing for conjecturing and proving: Developing principles based on practical tasks. In M. d. Villiers & G. Hanna (Eds.), *Proof and proving in mathematics education, ICME Study 19*: Springer.

Mason, J., Burton, L., & Stacey, K. (1982). *Thinking mathematically.* London and Reading, Mass.: Addison-Wesley Pub. Co.

Schön, D. (1983). *The reflective practitioner. How professionals think in action.* London: Basic Books.

Shulman, L. S. (1986). Those who understand: Knowledge growth in teaching. *Educational Researcher*, 4–14.

Vygotsky, L. (1978). *Mind in society: The development of higher psychological processes.* Cambridge, MA: Harvard University Press. Published originally in Russian in 1930.

Weimer, M. (2013). *Learner-centered teaching: Five key changes to practice* (2nd ed.). San Francisco: Jossey-Bass.

Zaslavsky, O. (2008). Meeting the challenges of mathematics teacher education through design and use of tasks that facilitate teacher learning. In B. Jaworski & T. Wood (Eds.), *The mathematics teacher educator as a developing professional* (pp. 93–114). Rotterdam, the Netherlands: Sense Publishers.

Zaslavsky, O., & Leikin, R. (2004). Professional development of mathematics teacher educators: Growth through practice. *Journal of Mathematics Teacher Education, 7*, 5–32.

FSC
www.fsc.org
MIX
Papier | Fördert gute Waldnutzung
FSC® C083411

Kolibri 360 GmbH
Ferdinand-Jühlke-Straße 7
99095 Erfurt, Deutschland
produktsicherheit@kolibri360.de